SCUOLA NORMALE SUPERIORE

CATTEDRA GALILEIANA

Walter Schachermayer

Portfolio Optimization in Incomplete Financial Markets

PISA 2004

Walter Schachermayer
Vienna University of Technology
Wiedner Hauptstrasse 8-10
1040 Vienna, Austria

Portfolio Optimization in Incomplete Financial Markets

Contents

Foreword

These Lecture Notes are based on a course given in June 2001 at the Cattedra Galileiano of Scuola Normale Superiore di Pisa. The course consisted in a short introduction into the basic concepts of Mathematical Finance, focusing on the notion of "no arbitrage", and subsequently applying these concepts to portfolio optimisation. To avoid technical difficulties I mainly dealt with the situation where the underlying probability space $(\Omega, \mathcal{F}, \mathbf{P})$ is finite and only sketched the difficulties arising in the general case. This part of the lectures is strongly based on my lecture notes for the summer school in St. Flour [S03] and the survey given at the first world congress of the Bachelier society [S01a].

We then pass to the theme of utility optimisation for general semi-martingale models as developped in [KS99] and [S01].

There are, however, some topics of this course which are not standard: for example, in the treatment of the general existence theorem for the optimal portfolio, we give a direct proof which is not relying on duality theory. Similarly, the treatment of the *asymptotic elasticity* of utility functions and a related counter-example are original to these notes, who were taken by P. Guasoni. He gave these notes his personal flavor, in particular with respect to these novel features: for example, he pointed out to me an interesting connection between the elasticity and the relative risk aversion of the utility function U. Our discussions on these topics of the course also resulted in the subsequent joint paper [GS04].

My sincerest thanks go to P. Guasoni for his dedication to the cumbersome duty of writing up my lectures and bringing them into shape. My thanks also go to Professors Pratelli and da Prato who were lovely hosts during my stay in Pisa.

Vienna, January 2004

Walter Schachermayer

CHAPTER I

Problem Setting

We consider a model of a security market which consists of $d+1$ assets. We denote by $S = ((S_t^i)_{1 \leq t \leq T})_{0 \leq i \leq d}$ the price process of the d stocks and suppose that the price of the asset S^0, called the "bond" or "cash account", is constant, i.e., $S_t^0 \equiv 1$. The latter assumption does not restrict the generality of the model as we always may choose the bond as numéraire, i.e., we may express the values of the other assets in units of the "bond". In other words, $((S_t^i)_{0 \leq t \leq T})_{1 \leq i \leq d}$, is an $\mathbb{R}^d$-valued semi-martingale modeling the discounted price process of d risky assets.

The process S is assumed to be a semimartingale, based on and adapted to a filtered probability space $(\Omega, \mathcal{F}, (\mathcal{F}_t)_{0 \leq t \leq T}, \mathbf{P})$ satisfying the usual conditions of saturatedness and right continuity. As usual in mathematical finance, we consider a finite horizon T, but we remark that our results can also be extended to the case of an infinite horizon.

In Chapter 2 we shall consider the case of finite Ω, in which case the paths of S are constant except for jumps at a finite number of times. We then can write S as $(S_t)_{t=0}^T = (S_0, S_1, \ldots, S_T)$, for some $T \in \mathbb{N}$.

The assumption that the bond is constant is mainly chosen for notational convenience as it allows for a compact description of self-financing portfolios: a self-financing portfolio Π is defined as a pair (x, H), where the constant x is the initial value of the portfolio and $H = (H^i)_{1 \leq i \leq d}$ is a predictable S-integrable process specifying the amount of each asset held in the portfolio. The value process $X = (X_t)_{0 \leq t \leq T}$ of such a portfolio Π at time t is given by

$$(1.1) \qquad X_t = X_0 + \int_0^t H_u \, dS_u, \quad 0 \leq t \leq T,$$

where $X_0 = x$ and the integral refers to stochastic integration in $\mathbb{R}^d$.

In order to rule out doubling strategies and similar schemes generating arbitrage-profits (by going deeply into the red) we follow Harrison and Pliska ([HP81], see also [DS94]), calling a predictable, S-integrable process *admissible*, if there is a constant $C \in \mathbb{R}_+$ such that, almost surely, we have

$$(1.2) \qquad (H \cdot S)_t := \int_0^t H_u dS_u \geq -C, \qquad \text{for } 0 \leq t \leq T.$$

Let us illustrate these general concepts in the case of an $\mathbb{R}^d$-valued process $S = (S_t)_{t=0}^T$ in finite, discrete time $\{0, 1, \dots, T\}$ adapted to the filtration $(\mathcal{F}_t)_{t=0}^T$. In this case each $\mathbb{R}^d$-valued process $(H_t)_{t=1}^T$, which is predictable (i.e. each H_t is $\mathcal{F}_{t-1}$-measurable), is S-integrable, and the stochastic integral reduces to a finite sum

$$(1.3) \qquad (H \cdot S)_t = \int_0^t H_u dS_u$$

$$(1.4) \qquad\qquad\quad = \sum_{u=1}^t H_u \Delta S_u$$

$$(1.5) \qquad\qquad\quad = \sum_{u=1}^t H_u (S_u - S_{u-1}),$$

where $H_u \Delta S_u$ denotes the inner product of the vectors H_u and $\Delta S_u = S_u - S_{u-1}$ in $\mathbb{R}^d$, i.e.

$$(1.6) \qquad H_u \Delta S_u = \sum_{j=1}^d H_u^j (S_u^j - S_u^{j-1}).$$

Of course, each such trading strategy H is admissible if the underlying probability space Ω is finite.

Passing again to the general setting of an $\mathbb{R}^d$-valued semi-martingale $S = (S_t)_{0 \leq t \leq T}$ we denote as in [KS99] by $\mathcal{M}^e(S)$ (resp $\mathcal{M}^a(S)$) the set of probability measures $\mathbf{Q}$ equivalent to $\mathbf{P}$ (resp. absolutely continuous with respect to $\mathbf{P}$) such that for each admissible integrand H, the process $H \cdot S$ is a local martingale under $\mathbf{Q}$.

We shall assume the following version of the no-arbitrage condition on S:

ASSUMPTION 1.1. The set $\mathcal{M}^e(S)$ is not empty.[1]

---

[1]If follows from [DS94] and [DS98] that Assumption 1.1 is equivalent to the condition of "no free lunch with vanishing risk". This property can also be equivalently characterised in terms of the existence of a measure $\mathbf{Q} \sim \mathbf{P}$ such that the process S itself (rather than the integrals $H \cdot S$ for admissible integrands) is "something like a martingale". The precise notion in the general semi-martingale setting is that S is a sigma-martingale under $\mathbf{Q}$ (see [DS98]); in the case when S is locally bounded (resp. bounded) the term "sigma-martingale" may be replaced by the more familiar term "local martingale" (resp. "martingale").

Readers who are not too enthusiastic about the rather subtle distinctions between martingales, local martingales and sigma-martingales may find some relief by noting that, in the case of finite Ω, or, more generally, for bounded processes S, these three notions coincide.

In these notes we shall mainly be interested in the case when $\mathcal{M}^e(S)$ is not reduced to a singleton, i.e., the case of an *incomplete* financial market.

After having specified the process S modeling the financial market we now define the function $U(x)$ modeling the utility of an agent's wealth x at the terminal time T.

We make the classical assumptions that $U : \mathbb{R} \to \mathbb{R} \cup \{-\infty\}$ is *increasing on $\mathbb{R}$, continuous* on $\{U > -\infty\}$, *differentiable and strictly concave* on the interior of $\{U > -\infty\}$, and that marginal utility tends to zero when wealth tends to infinity, i.e.,

$$(1.7) \qquad U'(\infty) := \lim_{x \to \infty} U'(x) = 0.$$

These assumptions make good sense economically and it is clear that the requirement (1.7) of marginal utility decreasing to zero, as x tends to infinity, is necessary, if one is aiming for a general existence theorem for optimal investment.

As regards the behavior of the (marginal) utility at the other end of the wealth scale we shall distinguish two cases.

CASE 1 (negative wealth not allowed). In this setting we assume that U satifies the conditions $U(x) = -\infty$, for $x < 0$, while $U(x) > -\infty$, for $x > 0$, and the so-called *Inada conditions*

$$(1.8) \qquad U'(0) := \lim_{x \searrow 0} U'(x) = \infty.$$

CASE 2 (negative wealth allowed). In this case we assume that $U(x) > -\infty$, for all $x \in \mathbb{R}$, and that

$$(1.9) \qquad U'(-\infty) := \lim_{x \searrow -\infty} U'(x) = \infty.$$

Typical examples for Case 1 are

$$(1.10) \qquad U(x) = \ln(x), \quad x > 0,$$

or

$$(1.11) \qquad U(x) = \frac{x^\alpha}{\alpha}, \quad \alpha \in (-\infty, 1) \setminus \{0\}, \quad x > 0,$$

whereas a typical example for Case 2 is

$$(1.12) \qquad U(x) = -e^{-\gamma x}, \quad \gamma > 0, \quad x \in \mathbb{R}.$$

We again note that it is natural from economic considerations to require that the marginal utility tends to infinity when the wealth x tends to the infimum of its allowed values.

For later reference we summarize our assumptions on the utility function:

ASSUMPTION 1.2 (Usual Regularity Conditions). A utility function $U : \mathbb{R} \to \mathbb{R} \cup \{-\infty\}$ satisfies the *usual regularity conditions* if it is increasing on $\mathbb{R}$, continuous on $\{U > -\infty\}$, differentiable and strictly concave on the interior of $\{U > -\infty\}$, and satisfies

$$(1.13) \qquad U'(\infty) := \lim_{x \to \infty} U'(x) = 0.$$

Denoting by dom(U) the interior of $\{U > -\infty\}$, we assume that we have one of the two following cases.

CASE 1. dom(U) $=]0, \infty[$ in which case U satisfies the condition

$$(1.14) \qquad U'(0) := \lim_{x \searrow 0} U'(x) = \infty.$$

CASE 2. dom(U) $= \mathbb{R}$ in which case U satisfies

$$(1.15) \qquad U'(-\infty) := \lim_{x \searrow -\infty} U'(x) = \infty.$$

We now can give a precise meaning to the problem of maximizing the expected utility of terminal wealth. Define the value function

$$(1.16) \qquad u(x) := \sup_{H \in \mathcal{H}} \mathbf{E}\big[U(x + (H \cdot S)_T)\big], \quad x \in \text{dom}(U),$$

where H ranges through the family $\mathcal{H}$ of admissible S-integrable trading strategies. To exclude trivial cases we shall assume that the value function u is not degenerate:

ASSUMPTION 1.3.

$$(1.17) \qquad u(x) < \sup_{\xi} U(\xi), \quad \text{for some} \quad x \in \text{dom}(U).$$

Since u is clearly increasing, and $U(y) \leq U(x) + U'(x)(y - x)$ for any $y > x$, this assumption implies that

$$(1.18) \qquad u(x) < \sup_{\xi} U(\xi), \quad \text{for all} \quad x \in \text{dom}(U).$$

Under appropriate hypotheses (e.g., when Ω is finite) Assumptions 1.1 and 1.2 already imply Assumption 1.3.

CHAPTER II

Models on Finite Probability Spaces

In order to reduce the technical difficulties of the theory of utility maximization to a minimum, we assume throughout this chapter that the probability space Ω will be finite, say, $\Omega = \{\omega_1, \omega_2, \dots, \omega_N\}$. This assumption implies that all the differences among the spaces $L^\infty(\Omega, \mathcal{F}, \mathbf{P})$, $L^1(\Omega, \mathcal{F}, \mathbf{P})$ and $L^0(\Omega, \mathcal{F}, \mathbf{P})$ disappear, as all these spaces are simply isomorphic to $\mathbb{R}^N$. Hence all the functional analysis reduces to simple linear algebra in the setting of the present chapter.

Nevertheless we shall write $L^\infty(\Omega, \mathcal{F}, \mathbf{P})$, $L^1(\Omega, \mathcal{F}, \mathbf{P})$ etc. below (knowing very well that these spaces are isomorphic in the present setting) to indicate, what we shall encounter in the setting of the general theory.

DEFINITION 2.1. A model of a *finite financial market* is an $\mathbb{R}^{d+1}$-valued stochastic process $S = (S)_{t=0}^T = (S_t^0, S_t^1, \dots, S_t^d)_{t=0}^T$, based on and adapted to the filtered stochastic base $(\Omega, \mathcal{F}, (\mathcal{F})_{t=0}^T, \mathbf{P})$. Without loss of generality we assume that $\mathcal{F}_0$ is trivial, that $\mathcal{F}_T = \mathcal{F}$ is the power set of Ω, and that $\mathbf{P}[\omega_n] > 0$, for all $1 \le n \le N$. We assume that the zero coordinate S^0, which we call the cash account, satisfies $S_t^0 \equiv 1$, for $t = 0, 1, \dots, T$. The letter ΔS_t denotes the increment $S_t - S_{t-1}$.

DEFINITION 2.2. $\mathcal{H}$ denotes the set of *trading strategies* for the financial market S. An element $H \in \mathcal{H}$ is an $\mathbb{R}^d$-valued process $(H_t)_{t=1}^T = (H_t^1, H_t^2, \dots, H_t^d)_{t=1}^T$ which is predictable, i.e. each H_t is $\mathcal{F}_{t-1}$-measurable.

We then define the stochastic integral $(H \cdot S)$ as the $\mathbb{R}$-valued process $((H \cdot S)_t)_{t=0}^T$ given by

$$(2.1) \qquad (H \cdot S)_t = \sum_{k=1}^t (H_k, \Delta S_k), \quad t = 0, \dots, T,$$

where $(.,.)$ denotes the inner product in $\mathbb{R}^d$.

DEFINITION 2.3. We call the subspace K of $L^0(\Omega, \mathcal{F}, \mathbf{P})$ defined by

$$(2.2) \qquad\qquad K = \{(H \cdot S)_T : H \in \mathcal{H}\}$$

the *set of contingent claims attainable at price 0.*

The economic interpretation is the following: the random variables $f = (H \cdot S)_T$, for some $H \in \mathcal{H}$, are precisely those contingent claims, i.e., the pay-off functions at time T depending on $\omega \in \Omega$ in an $\mathcal{F}_T$-measurable way, that an economic agent may replicate with zero initial investment, by pursuing some predictable trading strategy H.

For $a \in \mathbb{R}$, we call the *set of contingent claims attainable at price a* the affine space K_a obtained by shifting K by the constant function $a\mathbf{1}$, in other words the random variables of the form $a + (H \cdot S)_T$, for some trading strategy H. Again the economic interpretation is that these are precisely the contingent claims that an economic agent may replicate with an initial investment of a by pursuing some predictable trading strategy H.

DEFINITION 2.4. We call the convex cone C in $L^\infty(\Omega, \mathcal{F}, \mathbf{P})$ defined by

$$(2.3) \qquad C = \{g \in L^\infty(\Omega, \mathcal{F}, \mathbf{P}) \ \text{ s.t. there is } \ f \in K, f \geq g\}$$

the *set of contingent claims super-replicable at price 0.*

Economically speaking, a contingent claim $g \in L^\infty(\Omega, \mathcal{F}, \mathbf{P})$ is *super-replicable at price* 0, if we can achieve it with zero net investment, subsequently pursuing some predictable trading strategy H — thus arriving at some contingent claim f — and then, possibly, "throwing away money" to arrive at g. This operation of "throwing away money" may seem awkward at this stage, but we shall see later that the set C plays an important role in the development of the theory. Observe that C is a convex cone containing the negative orthant $L^\infty_-(\Omega, \mathcal{F}, \mathbf{P})$. Again we may define C_a as the *contingent claims super-replicable at price a* if we shift C by the constant function $a\mathbf{1}$.

DEFINITION 2.5. A financial market S satifies the *no-arbitrage condition (NA)* if

$$(2.4) \qquad\qquad K \cap L^0_+(\Omega, \mathcal{F}, \mathbf{P}) = \{0\}$$

or, equivalently,

$$(2.5) \qquad\qquad C \cap L^\infty_+(\Omega, \mathcal{F}, \mathbf{P}) = \{0\}$$

where 0 denotes the function identically equal to zero.

In other words we now have formalized the concept of an arbitrage possibility: it consists of the existence of a trading strategy H such that — starting from an initial investment zero — the resulting contingent claim $f = (H \cdot S)_T$ is non-negative and not identically equal to zero. If a financial market does not allow for arbitrage we say it satisfies the *no-arbitrage condition (NA)*.

DEFINITION 2.6. A probability measure $\mathbf{Q}$ on $(\Omega, \mathcal{F})$ is called an *equivalent martingale measure* for S, if $\mathbf{Q} \sim \mathbf{P}$ and S is a martingale under $\mathbf{Q}$.

We denote by $\mathcal{M}^e(S)$ the set of equivalent martingale probability measures and by $\mathcal{M}^a(S)$ the set of all (not necessarily equivalent) martingale probability measures. The letter a stands for "absolutely continuous with respect to $\mathbf{P}$" which in the present setting (finite Ω and $\mathbf{P}$ having full support) automatically holds true, but which will be of relevance for general probability spaces $(\Omega, \mathcal{F}, \mathbf{P})$ later. We shall often identify a measure $\mathbf{Q}$ on $(\Omega, \mathcal{F})$ with its Radon-Nikodym derivative $\frac{d\mathbf{Q}}{d\mathbf{P}} \in L^1(\Omega, \mathcal{F}, \mathbf{P})$.

LEMMA 2.7. *For a probability measure $\mathbf{Q}$ on $(\Omega, \mathcal{F})$ the following are equivalent:*
(i) $\mathbf{Q} \in \mathcal{M}^a(S)$,
(ii) $\mathbf{E}_{\mathbf{Q}}[f] = 0$, *for all* $f \in K$,
(iii) $\mathbf{E}_{\mathbf{Q}}[g] \leq 0$, *for all* $g \in C$.

PROOF. The equivalences are rather trivial, as (ii) is tantamount to the very definition of S being a martingale under $\mathbf{Q}$, and the equivalence of (ii) and (iii) is straightforward. □

After having fixed these formalities we may formulate and prove the central result of the theory of pricing and hedging by no-arbitrage, sometimes called the "fundamental theorem of asset pricing", which in its present form (i.e., finite Ω) is due to Harrison and Pliska [HP81].

THEOREM (Fundamental Theorem of Asset Pricing). *For a financial market S modeled on a finite stochastic base $(\Omega, \mathcal{F}, (\mathcal{F}_t)_{t=0}^T, \mathbf{P})$ the following are equivalent:*
(i) S *satisfies (NA).*
(ii) $\mathcal{M}^e(S) \neq \emptyset$.

PROOF. (ii) $\Rightarrow$ (i): this is the obvious implication. If there is some $\mathbf{Q} \in \mathcal{M}^e(S)$ then by Lemma 2.7 we have that

$$(2.6) \qquad \mathbf{E}_{\mathbf{Q}}[g] \leq 0, \quad \text{for} \quad g \in C.$$

On the other hand, if there were $g \in C \cap L_+^\infty$, $g \neq 0$, then, using the assumption that $\mathbf{Q}$ is equivalent to $\mathbf{P}$, we would have

$$(2.7) \qquad \mathbf{E}_{\mathbf{Q}}[g] > 0,$$

a contradiction.

(i) $\Rightarrow$ (ii): this implication is the important message of the theorem which will allow us to link the no-arbitrage arguments with martingale theory. We give a functional analytic existence proof, which will be generalizable — in spirit — to more general situations.

By assumption the space K intersects L^∞_+ only at 0. We want to separate the disjoint convex sets $L^\infty_+\backslash\{0\}$ and K by a hyperplane induced by a linear functional $\mathbf{Q} \in L^1(\Omega, \mathcal{F}, \mathbf{P})$ which is *strictly positive* on $L^\infty_+\backslash\{0\}$. Unfortunately this is a situation, where the usual versions of the separation theorem (i.e., the Hahn-Banach theorem) do not apply (even in finite dimensions!). Indeed, one usually assumes that one of the convex sets is compact in order to obtain a strict separation.

One way to overcome this difficulty (in finite dimension) is to consider the convex hull of the unit vectors $(\mathbf{1}_{\{\omega_n\}})_{n=1}^N$ in $L^\infty(\Omega, \mathcal{F}, \mathbf{P})$ i.e.

$$(2.8) \qquad P := \left\{ \sum_{n=1}^N \mu_n \mathbf{1}_{\{\omega_n\}} : \mu_n \geq 0, \sum_{n=1}^N \mu_n = 1 \right\}.$$

This is a convex, compact subset of $L^\infty_+(\Omega, \mathcal{F}, \mathbf{P})$ and, by the (*NA*) assumption, disjoint from K. Hence we may strictly separate the sets P and K by a linear functional $\mathbf{Q} \in L^\infty(\Omega, \mathcal{F}, \mathbf{P})^* = L^1(\Omega, \mathcal{F}, \mathbf{P})$, i.e., find $\alpha < \beta$ such that

$$(2.9) \qquad \begin{aligned} \mathbf{E_Q}[f] &= \langle \mathbf{Q}, f \rangle \leq \alpha \quad \text{for} \quad f \in K, \\ \langle \mathbf{Q}, h\mathbf{E_Q}[h] =\rangle &\geq \beta \quad \text{for} \quad h \in P. \end{aligned}$$

As K is a linear space, we have $\alpha \geq 0$ and may, in fact, replace α by 0. Hence $\beta > 0$. Therefore $\langle \mathbf{Q}, \mathbf{1} \rangle > 0$, and we may normalize $\mathbf{Q}$ such that $\langle \mathbf{Q}, \mathbf{1} \rangle = 1$. As $\mathbf{Q}$ is strictly positive on each $\mathbf{1}_{\{\omega_n\}}$, we therefore have found a probability measure $\mathbf{Q}$ on $(\Omega, \mathcal{F})$ equivalent to $\mathbf{P}$ such that condition (ii) of Lemma 2.7 holds true. In other words, we found an equivalent martingale measure $\mathbf{Q}$ for the process S. □

COROLLARY 2.9. *Let S satisfy (NA) and $f \in L^\infty(\Omega, \mathcal{F}, \mathbf{P})$ be an attainable contingent claim so that*

$$(2.10) \qquad f = a + (H \cdot S)_T,$$

for some $a \in \mathbb{R}$ and some trading strategy H.

Then the constant a and the process $(H \cdot S)$ are uniquely determined by (2.10) and satisfy, for every $\mathbf{Q} \in \mathcal{M}^e(S)$,

$$(2.11) \qquad a = \mathbf{E_Q}[f], \quad \text{and} \quad a + (H \cdot S)_t = \mathbf{E_Q}[f|\mathcal{F}_t] \quad \text{for} \quad 0 \leq t \leq T.$$

PROOF. As regards the uniqueness of the constant $a \in \mathbb{R}$, suppose that there are two representations $f = a^1 + (H^1 \cdot S)_T$ and $f = a^2 + (H^2 \cdot S)_T$ with $a^1 \neq a^2$. Assuming w.l.o.g. that $a^1 > a^2$ we find an obvious arbitrage possibility: we have $a^1 - a^2 = ((H^1 - H^2) \cdot S)_T$, i.e. the trading strategy $H^1 - H^2$ produces a strictly positive result at time T, a contradiction to (*NA*).

As regards the uniqueness, or the process $H{\cdot}S$ we simply apply a conditional version of the previous argument: assume that $f = a + (H^1 \cdot S)_T$ and $f = a + (H^2 \cdot S)_T$ such that the processes $H^1{\cdot}S$ amd $H^2{\cdot}S$ are not identical. Then there is $0 < t < T$ such that $(H^1 \cdot S)_t \neq (H^2 \cdot S)_t$; w.l.g. $A := \{(H^1 \cdot S)_t > (H^2 \cdot S)_t\}$ is a non-empty event, which clearly is in $\mathcal{F}_t$. Hence, using the fact hat $(H^1 \cdot S)_T = (H^2 \cdot S)_T$, the trading strategy $H := (H^2 - H^1)\chi_A \cdot \chi_{]t,T]}$ is a predictable process producing an arbitrage, as $(H \cdot S)_T = 0$ outside A, while $(H \cdot S)_T = (H^1 \cdot S)_t - (H^2 \cdot S)_t > 0$ on A, which again contradicts (NA).

Finally, the equations in (2.11) result from the fact that, for every predictable process H and every $\mathbf{Q} \in \mathcal{M}^a(S)$, the process $H \cdot S$ is a $\mathbf{Q}$-martingale. Noting that, for a measure $\mathbf{Q} \sim \mathbf{P}$, the conditional expectation $\mathbf{E}_\mathbf{Q}[f|\mathcal{F}_t]$ is $\mathbf{P}$-a.s. well-defined we thus obtain (2.11) for each $\mathbf{Q} \in \mathcal{M}^e(S)$. $\square$

Denote by $\text{cone}(\mathcal{M}^e(S))$ and $\text{cone}(\mathcal{M}^a(S))$ the cones generated by the convex sets $\mathcal{M}^e(S)$ and $\mathcal{M}^a(S)$ respectively. The subsequent result clarifies the polar relation between these cones and the cone C. Recall (see, e.g., [S66]) that, for a pair (E, E') of vector spaces in separating duality via the scalar product $\langle ., . \rangle$, the polar C^0 of a set C in E is defined as

$$(2.12) \qquad C^0 = \{g \in E' : \langle f, g \rangle \leq 1, \quad \text{for all} \quad f \in C\}.$$

In the case when C is closed under multiplication with positive scalars (e.g., if C is a convex cone) the polar C^0 may equivalently be defined by

$$(2.13) \qquad C^0 = \{g \in E' : \langle f, g \rangle \leq 0, \quad \text{for all} \quad f \in C\}.$$

The *bipolar theorem* (see, e.g., [S66]) states that the bipolar $C^{00} := (C^0)^0$ of a set C in E is the $\sigma(E, E')$-closed convex hull of C.

After these general considerations we pass to the concrete setting of the cone $C \subseteq L^\infty(\Omega, \mathcal{F}, \mathbf{P})$ of contingent claims super-replicable at price 0. Note that in our finite-dimensional setting this convex cone is closed as it is the algebraic sum of the closed linear space K (a linear space in $\mathbb{R}^N$ is always closed) and the closed polyhedral cone $L^\infty_-(\Omega, \mathcal{F}, \mathbf{P})$ (the verification, that the algebraic sum of a space and a polyhedral cone in $\mathbb{R}^N$ is closed, is an easy, but not completely trivial exercise). Hence we deduce from the bipolar theorem, that C equals its bipolar C^{00}.

PROPOSITION 2.10. *Suppose that S satisfies (NA). Then the polar of C is equal to* $\text{cone}(\mathcal{M}^a(S))$ *and $\mathcal{M}^e(S)$ is dense in $\mathcal{M}^a(S)$. Hence the following assertions are equivalent for an element $g \in L^\infty(\Omega, \mathcal{F}, \mathbf{P})$*

(i) $g \in C$,
(ii) $\mathbf{E}_\mathbf{Q}[g] \leq 0$, *for all $g \in \mathcal{M}^a(S)$,*
(iii) $\mathbf{E}_\mathbf{Q}[g] \leq 0$, *for all $g \in \mathcal{M}^e(S)$,*

PROOF. The fact that the polar C^0 and $\text{cone}(\mathcal{M}^a(S))$ coincide, follows from Lemma 2.7 and the observation that $C \supseteq L^\infty_-(\Omega, \mathcal{F}, \mathbf{P})$ implies $C^0 \subseteq L^\infty_+(\Omega, \mathcal{F}, \mathbf{P})$. Hence the equivalence of (i) and (ii) follows from the bipolar theorem.

As regards the density of $\mathcal{M}^e(S)$ in $\mathcal{M}^a(S)$ we first deduce from Theorem 2.8 that there is at least one $\mathbf{Q}^* \in \mathcal{M}^e(S)$. For any $\mathbf{Q} \in \mathcal{M}^a(S)$ and $0 < \mu \leq 1$ we have that $\mu\mathbf{Q}^* + (1 - \mu)\mathbf{Q} \in \mathcal{M}^e(S)$, which clearly implies the density of $\mathcal{M}^e(S)$ in $\mathcal{M}^a(S)$. The equivalence of (ii) and (iii) now is obvious. $\square$

The subsequent theorem tells us precisely what the principle of no arbitrage can tell us about the possible prices for a contingent claim f. It goes back to the work of D. Kreps [K81] and was subsequently extended by several authors.

For given $f \in L^\infty(\Omega, \mathcal{F}, \mathbf{P})$, we call $a \in \mathbb{R}$ an *arbitrage-free price*, if in addition to the financial market S, the introduction of the contingent claim, which pays the random amount f at time $t = T$ and can be bought or sold at price a at time $t = 0$, does not create an arbitrage possibility. Mathematically speaking, this can be formalized as follows. Let $C^{f,a}$ denote the cone spanned by C and the linear space spanned by $f - a$; then a is an arbitrage-free price for f if $C^{f,a} \cap L^\infty_+(\Omega, \mathcal{F}, \mathbf{P}) = \{0\}$.

THEOREM 2.11 (Pricing by No-Arbitrage). *Assume that S satisfies* (NA) *and let $f \in L^\infty(\Omega, \mathcal{F}, \mathbf{P})$. Define*

(2.14) $$\overline{\pi}(f) = \sup\{\mathbf{E}_Q[f] : Q \in \mathcal{M}^e(S)\},$$

(2.15) $$\underline{\pi}(f) = \inf\{\mathbf{E}_Q[f] : Q \in \mathcal{M}^e(S)\}.$$

Either $\underline{\pi}(f) = \overline{\pi}(f)$, in which case f is attainable at price $\pi(f) := \underline{\pi}(f) = \overline{\pi}(f)$, i.e. $f = \pi(f) + (H \cdot S)_T$ for some $H \in \mathcal{H}$; therefore $\pi(f)$ is the unique arbitrage-free price for f.

Or $\underline{\pi}(f) < \overline{\pi}(f)$, in which case $\{\mathbf{E}_Q[f] : Q \in \mathcal{M}^e(S)\}$ equals the open interval $]\underline{\pi}(f), \overline{\pi}(f)[$, which in turn equals the set of arbitrage-free prices for the contingent claim f.

PROOF. First observe that the set $\{\mathbf{E}_Q[f] : Q \in \mathcal{M}^e(S)\}$ forms a bounded non-empty interval in $\mathbb{R}$, which we denote by I.

We claim that a number a is in I, iff a is an arbitrage-free price for f. Indeed, supposing that $a \in I$ we may find $Q \in \mathcal{M}^e(S)$ s.t. $\mathbf{E}_Q[f - a] = 0$ and therefore $C^{f,a} \cap L^\infty_+(\Omega, \mathcal{F}, \mathbf{P}) = \{0\}$.

Conversely suppose that $C^{f,a} \cap L^\infty_+(\Omega, \mathcal{F}, \mathbf{P}) = \{0\}$. Note that $C^{f,a}$ is a closed convex cone (it is the albegraic sum of the linear space $\mathrm{span}(K, f - a)$ and the closed, polyhedral cone $L^\infty_-(\Omega, \mathcal{F}, \mathbf{P})$). Hence by the same argument as in the proof of Theorem 2.8 there exists a probability measure $Q \sim \mathbf{P}$ such that $Q|_{C^{f,a}} \leq 0$. This implies that $\mathbf{E}_Q[f - a] = 0$, i.e., $a \in I$.

Now we deal with the boundary case: suppose that a equals the right boundary of I, i.e., $a = \overline{\pi}(f) \in I$, and consider the contingent claim $f - \overline{\pi}(f)$; by definition we have $\mathbf{E}_Q[f - \overline{\pi}(f)] \leq 0$, for all $Q \in \mathcal{M}^e(S)$, and therefore by Proposition 2.10, that $f - \overline{\pi}(f) \in C$. We may find $g \in K$ such that $g \geq f - \overline{\pi}(f)$. If the sup in (2.14) is attained, i.e., if there is $Q^* \in \mathcal{M}^e(S)$ such that $\mathbf{E}_{Q^*}[f] = \overline{\pi}(f)$, then we have $0 = \mathbf{E}_{Q^*}[g] \geq \mathbf{E}_{Q^*}[f - \overline{\pi}(f)] = 0$ which in view of $Q^* \sim \mathbf{P}$ implies that $f - \overline{\pi}(f) \equiv g$; in other words f is attainable at price $\overline{\pi}(f)$. This in turn implies that $\mathbf{E}_Q[f] = \overline{\pi}(f)$, for all $Q \in \mathcal{M}^e(S)$, and therefore I is reduced to the singleton $\{\overline{\pi}(f)\}$.

Hence, if $\underline{\pi}(f) < \overline{\pi}(f)$, $\overline{\pi}(f)$ connot belong to the interval I, which is therefore open on the right hand side. Passing from f to $-f$, we obtain the analogous result for the left hand side of I, which therefore equals $I =]\underline{\pi}(f), \overline{\pi}(f)[$.   □

COROLLARY 2.12 (complete financial markets). *For a financial market S satisfying the no-arbitrage condition (NA) the following are equivalent*:
  (i) $\mathcal{M}^e(S)$ *consists of a single element* **Q**.
 (ii) *Each* $f \in L^\infty(\Omega, \mathcal{F}, \mathbf{P})$ *may be represented as*

(2.16) $$f = a + (H \cdot S)_T, \quad \text{for some} \quad a \in \mathbb{R}, \quad \text{and} \quad H \in \mathcal{H}.$$

In this case $a = \mathbf{E}_\mathbf{Q}[f]$, *the stochastic integral* $(H \cdot S)$ *is unique and we have that*

(2.17) $$\mathbf{E}_\mathbf{Q}[f|\mathcal{F}_t] = \mathbf{E}_\mathbf{Q}[f] + (H \cdot S)_t, \quad t = 0, \dots, T.$$

PROOF. The implication (i) $\Rightarrow$ (ii) immediately follows from the preceding theorem; for the implication (ii) $\Rightarrow$ (i), note that, (2.16) implies that, for elements $\mathbf{Q}_1, \mathbf{Q}_2 \in \mathcal{M}^a(S)$, we have $\mathbf{E}_{\mathbf{Q}_1}[f] = a = \mathbf{E}_{\mathbf{Q}_2}[f]$; hence it suffices to note that, if $\mathcal{M}^e(S)$ contains two different elements $\mathbf{Q}_1, \mathbf{Q}_2$, we may find $f \in L^\infty(\Omega, \mathcal{F}, \mathbf{P})$ s.t. $\mathbf{E}_{\mathbf{Q}_1}[f] \neq \mathbf{E}_{\mathbf{Q}_2}[f]$. $\square$

2.1. – Utility maximization

We are now ready to study utility maximization problems with the convex duality approach.

2.1.1. – The complete case (Arrow)

As a first case we analyze the situation of a *complete* financial market (Corollary 2.12 above), i.e., the set $\mathcal{M}^e(S)$ of equivalent probability measures under which S is a martingale is reduced to a singleton $\{\mathbf{Q}\}$. In this setting consider the *Arrow assets* $\mathbf{1}_{\{\omega_n\}}$, which pay 1 unit of the numéraire at time T, when ω_n turns out to be the true state of the world, and 0 otherwise. In view of our normalization of the numéraire $S_t^0 \equiv 1$, we get for the price of the Arrow assets at time $t = 0$ the relation

(2.18) $$\mathbf{E}_\mathbf{Q}\left[\mathbf{1}_{\{\omega_n\}}\right] = \mathbf{Q}[\omega_n] = q_n,$$

and by 2.12 each Arrow asset $\mathbf{1}_{\{\omega_n\}}$ may be represented as $\mathbf{1}_{\{\omega_n\}} = \mathbf{Q}[\omega_n] + (H \cdot S)_T$, for some predictable trading strategy $H \in \mathcal{H}$.
    Hence, for fixed initial endowment $x \in \text{dom}(U)$, the utility maximization problem (1.16) above may simply be written as

(2.19) $$\mathbf{E}_\mathbf{P}\left[U(X_T)\right] = \sum_{n=1}^{N} p_n U(\xi_n) \to \max!$$

(2.20) $$\mathbf{E}_\mathbf{Q}[X_T] = \sum_{n=1}^{N} q_n \xi_n \leq x.$$

To verify that (2.19) and (2.20) indeed are equivalent to the original problem (2.16) above (in the present finite, complete case), note that by Theorem 2.11 a random variable $(X_T(\omega_n))_{n=1}^{N} = (\xi_n)_{n=1}^{N}$ can be dominated by a random variable of the form $x + (H \cdot S)_T = x + \sum_{t=1}^{T} H_t \Delta S_t$ iff $\mathbf{E_Q}[X_T] = \sum_{n=1}^{N} q_n \xi_n \leq x$. This basic relation has a particularly evident interpretation in the present setting, as q_n is simply the price of the Arrow asset $\mathbf{1}_{\{\omega_n\}}$.

We have written ξ_n for $X_T(\omega_n)$ to stress that (2.19) simply is a concave maximization problem in $\mathbb{R}^N$ with one linear constraint. To solve it, we form the Lagrangian

$$(2.21) \qquad L(\xi_1, \ldots, \xi_N, y) = \sum_{n=1}^{N} p_n U(\xi_n) - y \left(\sum_{n=1}^{N} q_n \xi_n - x \right)$$

$$(2.22) \qquad = \sum_{n=1}^{N} p_n \left(U(\xi_n) - y \frac{q_n}{p_n} \xi_n \right) + yx.$$

We have used the letter $y \geq 0$ instead of the usual $\lambda \geq 0$ for the Lagrange multiplier; the reason is the dual relation between x and y which will become apparent in a moment.

Write

$$(2.23) \qquad \Phi(\xi_1, \ldots, \xi_N) = \inf_{y>0} L(\xi_1, \ldots, \xi_N, y), \quad \xi_n \in \text{dom}(U),$$

and

$$(2.24) \qquad \Psi(y) = \sup_{\xi_1, \ldots, \xi_N} L(\xi_1, \ldots, \xi_N, y), \quad y \geq 0.$$

Note that we have

$$(2.25) \qquad \sup_{\xi_1, \ldots, \xi_N} \Phi(\xi_1, \ldots, \xi_N) = \sup_{\substack{\xi_1, \ldots, \xi_N \\ \sum_{n=1}^{N} q_n \xi_n \leq x}} \sum_{n=1}^{N} p_n U(\xi_n) = u(x).$$

Indeed, if $(\xi_1, \ldots, \xi_N)$ is in the admissible region $\sum_{n=1}^{N} q_n \xi_n \leq x$ then $\Phi(\xi_1, \ldots, \xi_N) = L(\xi_1, \ldots, \xi_N, 0) = \sum_{n=1}^{N} p_n U(\xi_n)$. On the other hand, if $(\xi_1, \ldots, \xi_N)$ satisfies $\sum_{n=1}^{N} q_n \xi_n > x$, then by letting $y \to \infty$ in (2.23) we note that $\Phi(\xi_1, \ldots, \xi_N) = -\infty$.

As regards the function $\Psi(y)$ we make the following pleasant observation which is the basic reason for the efficiency of the duality approach: using the form (2.22) of the Lagrangian and fixing $y > 0$, the optimization problem appearing in (2.24) splits into N independent optimization problems over $\mathbb{R}$

$$(2.26) \qquad U(\xi_n) - y \frac{q_n}{p_n} \xi_n \to \max!, \quad \xi_n \in \mathbb{R}.$$

In fact, these one-dimensional optimization problems are of a very convenient form: recall (see, e.g., [R70], [ET76] or [KLSX91]) that, for a concave function $U : \mathbb{R} \to \mathbb{R} \cup \{-\infty\}$, the *conjugate function* V (which is just the Legendre-transform of $x \mapsto -U(-x)$) is defined by

$$(2.27) \qquad V(\eta) = \sup_{\xi \in \mathbb{R}} \left[U(\xi) - \eta \xi \right], \quad \eta > 0.$$

DEFINITION 2.13. We say that the function $V : \mathbb{R} \to \mathbb{R}$, conjugate to the function U, satisfies the *usual regularity assumptions*, if V is finitely valued, differentiable, strictly convex on $]0, \infty[$, and satisfies

$$(2.28) \qquad V'(0) := \lim_{y \searrow 0} V'(y) = -\infty.$$

As regards the behavior of V at infinity, we have to distinguish between case 1 and case 2 in Assumption 1.2 above:

$$(2.29) \qquad \text{case 1:} \quad \lim_{y \to \infty} V(y) = \lim_{x \to 0} U(x) \quad \text{and} \quad \lim_{y \to \infty} V'(y) = 0$$

$$(2.30) \qquad \text{case 2:} \quad \lim_{y \to \infty} V(y) = \infty \quad \text{and} \quad \lim_{y \to \infty} V'(y) = \infty.$$

We have the following wellknown fact (see [R70] or [ET76]).

PROPOSITION 2.14. *If U satisfies Assumption 1.2, then its conjugate function V satisfies the the inversion formula*

$$(2.31) \qquad U(\xi) = \inf_{\eta} \left[V(\eta) + \eta \xi \right], \quad \xi \in \mathrm{dom}(U)$$

and satisfies the regularity assumptions in Definition 2.13. In addition, $-V'(y)$ is the inverse function of $U'(x)$. Conversely, if V satisfies the regulatory assumptions of Definition 2.13, then U defined by (2.31) satisfies Assumption 1.2.

Following $[KLS87]$ we denote $-V' = I$ (for "inverse" function).

PROOF. It follows from Assumption 1.2 that V is finitely valued on $]0, \infty[$. Note that we have that

$$(2.32) \qquad U(x) \leq a + yx \quad \forall x \in \mathrm{dom}(U) \qquad \Longleftrightarrow \qquad V(y) \leq a$$

which implies the inversion formula above. In turn, this formula shows that V is the supremum of affine functions, and therefore convex. Since U is strictly concave and differentiable, the maximizer $\hat{\xi} = \xi(\mu)$ in (2.27) solves the first-order condition $U'(\xi(\eta)) = \eta$. Also, we have that U' is a continuous bijection between $\{U > -\infty\}$ and $\mathbb{R}_+$. This observation and the inversion formula show that V is both strictly convex, differentiable, and that $-V'$ is the inverse of U'. $\qquad \square$

REMARK 2.15. Of course, U' has a good economic interpretation as the *marginal utility* of an economic agent modeled by the utility function U.

Here are some concrete examples of pairs of conjugate functions:

$$U(x) = \ln(x), \ x > 0, \qquad V(y) = -\ln(y) - 1,$$

$$U(x) = -\frac{e^{-\gamma x}}{\gamma}, \ x \in \mathbb{R}, \quad V(y) = \frac{y}{\gamma}(\ln(y) - 1), \ \gamma > 0$$

$$U(x) = \frac{x^\alpha}{\alpha}, \ x > 0, \qquad V(y) = \frac{1-\alpha}{\alpha} y^{\frac{\alpha}{\alpha-1}}, \ \alpha \in (-\infty, 1) \setminus \{0\}.$$

We now apply these general facts about the Legendre transformation to calculate $\Psi(y)$. Using definition (2.27) of the conjugate function V and (2.22), formula (2.24) becomes

$$(2.33) \qquad \Psi(y) = \sum_{n=1}^{N} p_n V\left(y \frac{q_n}{p_n}\right) + yx$$

$$(2.34) \qquad\qquad = \mathbf{E_P}\left[V\left(y \frac{d\mathbf{Q}}{d\mathbf{P}}\right)\right] + yx.$$

Denoting by $v(y)$ the dual value function

$$(2.35) \qquad v(y) := \mathbf{E_P}\left[V\left(y \frac{d\mathbf{Q}}{d\mathbf{P}}\right)\right] = \sum_{n=1}^{N} p_n V\left(y \frac{q_n}{p_n}\right), \quad y > 0,$$

the function v has the same qualitative properties as the function V listed in Definition 2.13, since it is a convex combination of V calculated on linearly scaled arguments.

Hence by (2.28), (2.29), and (2.30) we find, for fixed $x \in \mathrm{dom}(U)$, a unique $\hat{y} = \hat{y}(x) > 0$ such that $v'(\hat{y}(x)) = -x$, which therefore is the unique minimizer to the dual problem

$$(2.36) \qquad \Psi(y) = \mathbf{E_P}\left[V\left(y \frac{d\mathbf{Q}}{d\mathbf{P}}\right)\right] + yx = \min!$$

Fixing the critical value $\hat{y}(x)$, the concave function

$$(2.37) \qquad (\xi_1, \dots, \xi_N) \mapsto L(\xi_1, \dots, \xi_N, \hat{y}(x))$$

defined in (2.22) assumes its unique maximum at the point $(\hat{\xi}_1, \dots, \hat{\xi}_N)$ satisfying

$$(2.38) \qquad U'(\hat{\xi}_n) = \hat{y}(x) \frac{q_n}{p_n} \quad \text{or, equivalently,} \quad \hat{\xi}_n = I\left(\hat{y}(x) \frac{q_n}{p_n}\right),$$

so that we have

(2.39) $$\inf_{y>0} \Psi(y) = \inf_{y>0} \left(v(y) + xy\right)$$

(2.40) $$= v(\widehat{y}(x)) + x\widehat{y}(x)$$

(2.41) $$= L(\widehat{\xi}_1, \dots, \widehat{\xi}_N, \widehat{y}(x)).$$

Note that $\widehat{\xi}_n$ are in dom(U), for $1 \leq n \leq N$, so that L is continuously differentiable at $(\widehat{\xi}_1, \dots, \widehat{\xi}_N, \widehat{y}(x))$, which implies that the gradient of L vanishes at $(\widehat{\xi}_1, \dots, \widehat{\xi}_N, \widehat{y}(x))$ and, in particular, that $\frac{\partial}{\partial y}L(\xi_1, \dots, \xi_N, y)|_{(\widehat{\xi}_1, \dots, \widehat{\xi}_N, \widehat{y}(x))} = 0$. Hence we infer from (2.21) and the fact that $\widehat{y}(x) > 0$ that the constraint (2.20) is binding, i.e.,

(2.42) $$\sum_{n=1}^{N} q_n \widehat{\xi}_n = x,$$

and that

(2.43) $$\sum_{n=1}^{N} p_n U(\widehat{\xi}_n) = L(\widehat{\xi}_1, \dots, \widehat{\xi}_N, \widehat{y}(x)).$$

In particular, we obtain that

(2.44) $$u(x) = \sum_{n=1}^{N} p_n U(\widehat{\xi}_n).$$

Indeed, the inequality $u(x) \geq \sum_{n=1}^{N} p_n U(\widehat{\xi}_n)$ follows from (2.42) and (2.25), while the reverse inequality follows from (2.43) and the fact that for all $\xi_1, \dots, \xi_N$ verifying the constraint (2.20)

(2.45) $$\sum_{n=1}^{N} p_n U(\xi_n) \leq L(\xi_1, \dots, \xi_N, \widehat{y}(x)) \leq L(\widehat{\xi}_1, \dots, \widehat{\xi}_N, \widehat{y}(x)).$$

We shall write $\widehat{X}_T(x) \in C(x)$ for the optimizer $\widehat{X}_T(x)(\omega_n) = \widehat{\xi}_n$, $n = 1, \dots, N$.

Combining (2.39), (2.43) and (2.44) we note that the value functions u and v are conjugate:

(2.46) $$\inf_{y>0} \left(v(y) + xy\right) = v(\widehat{y}(x)) + x\widehat{y}(x) = u(x), \quad x \in \text{dom}(U),$$

which, by Proposition 2.14 the remarks after equation (2.35), implies that u inherits the properties of U listed in Assumption 1.2. The relation $v'(\widehat{y}(x)) = -x$ which was used to define $\widehat{y}(x)$, therefore translates into

(2.47) $$u'(x) = \widehat{y}(x), \quad \text{for} \quad x \in \text{dom}(U).$$

Let us summarize what we have proved:

THEOREM 2.16 (finite Ω, complete market). *Let the financial market* $S = (S_t)_{t=0}^T$ *be defined over the finite filtered probability space* $(\Omega, \mathcal{F}, (\mathcal{F})_{t=0}^T, \mathbf{P})$ *and satisfy* $\mathcal{M}^e(S) = \{\mathbf{Q}\}$, *and let the utility function* U *satisfy Assumption 1.2.*

Denote by $u(x)$ *and* $v(y)$ *the value functions*

$$(2.48) \qquad u(x) = \sup_{X_T \in C(x)} \mathbf{E}[U(X_T)], \quad x \in \text{dom}(U),$$

$$(2.49) \qquad v(y) = \mathbf{E}\left[V\left(y\frac{d\mathbf{Q}}{d\mathbf{P}}\right)\right], \qquad y > 0.$$

We then have:

(i) *The value functions* $u(x)$ *and* $v(y)$ *are conjugate and* u *inherits the qualitative properties of* U *listed in Assumption 1.2.*

(ii) *The optimizer* $\widehat{X}_T(x)$ *in (2.48) exists, is unique and satisfies*

$$(2.50) \qquad \widehat{X}_T(x) = I\left(y\frac{d\mathbf{Q}}{d\mathbf{P}}\right), \quad \text{or, equivalently,} \quad y\frac{d\mathbf{Q}}{d\mathbf{P}} = U'(\widehat{X}_T(x)),$$

where $x \in \text{dom}(U)$ *and* $y > 0$ *are related via* $u'(x) = y$ *or, equivalently,* $x = -v'(y)$.

(iii) *The following formulae for* u' *and* v' *hold true:*

$$(2.51) \qquad u'(x) = \mathbf{E_P}[U'(\widehat{X}_T(x))], \quad v'(y) = \mathbf{E_Q}\left[V'\left(y\frac{d\mathbf{Q}}{d\mathbf{P}}\right)\right]$$

$$(2.52) \qquad \begin{aligned} xu'(x) &= \mathbf{E_P}\left[\widehat{X}_T(x)U'(\widehat{X}_T(x))\right], \\ yv'(y) &= \mathbf{E_P}\left[y\frac{d\mathbf{Q}}{d\mathbf{P}}V'\left(y\frac{d\mathbf{Q}}{d\mathbf{P}}\right)\right]. \end{aligned}$$

PROOF. Items (i) and (ii) have been shown in the preceding discussion, hence we only have to show (iii). The formulae for $v'(y)$ in (2.51) and (2.52) immediately follow by differentiating the relation

$$(2.53) \qquad v(y) = \mathbf{E_P}\left[V\left(y\frac{d\mathbf{Q}}{d\mathbf{P}}\right)\right] = \sum_{n=1}^N p_n V\left(y\frac{q_n}{p_n}\right).$$

Of course, the formula for v' in (2.52) is an obvious reformulation of the one in (2.51). But we write both of them to stress their symmetry with the formulae for $u'(x)$.

The formula for u' in (2.51) translates via the relations exhibited in (ii) into the identity

$$(2.54) \qquad y = \mathbf{E_P}\left[y\frac{d\mathbf{Q}}{d\mathbf{P}}\right],$$

while the formula for $u'(x)$ in (2.52) translates into

$$(2.55) \qquad v'(y)y = \mathbf{E_P}\left[V'\left(y\frac{d\mathbf{Q}}{d\mathbf{P}}\right)y\frac{d\mathbf{Q}}{d\mathbf{P}}\right],$$

which we just have seen to hold true. □

REMARK 2.17. Firstly, let us recall the economic interpretation of (2.50)

$$(2.56) \qquad U'\big(\widehat{X}_T(x)(\omega_n)\big) = y\frac{q_n}{p_n}, \quad n = 1, \dots, N.$$

This equality means that, in every possible state of the world ω_n, the *marginal utility* $U'(\widehat{X}_T(x)(\omega_n))$ of the wealth of an optimally investing agent at time T is *proportional to the ratio of the price q_n of the corresponding Arrow security $1_{\{\omega_n\}}$ and the probability of its success $p_n = \mathbf{P}[\omega_n]$. This basic relation was analyzed in the fundamental work of K. Arrow and allows for a convincing economic interpretation: considering for a moment the situation where this proportionality relation fails to hold true, one immediately deduces from a marginal variation argument that the investment of the agent cannot be optimal. Indeed, by investing a little more in the more favorable asset and a little less in the less favorable the economic agent can strictly increase expected utility under the same budget constraint. Hence for the optimal investment the proportionality must hold true. The above result also identifies the proportionality factor as $y = u'(x)$, where x is the initial endowment of the investor. This also allows for an economic interpretation.

Theorem 2.16 indicates an easy way to solve the utility maximization at hand: calculate $v(y)$ by (2.49), which reduces to a simple one-dimensional computation; once we know $v(y)$, the theorem provides easy formulae to calculate all the other quantities of interest, e.g., $\widehat{X}_T(x)$, $u(x)$, $u'(x)$ etc.

Another message of the above theorem is that the value function $x \mapsto u(x)$ may be viewed as a utility function as well, sharing all the qualitative features of the original utility function U. This makes sense economically, as the "indirect utility" function $u(x)$ denotes the expected utility at time T of an agent with initial endowment x, after having optimally invested in the financial market S.

Let us also give an economic interpretation of the formulae for $u'(x)$ in item (iii) along these lines: suppose the initial endowment x is varied to $x+h$, for some small real number h. The economic agent may use the additional endowment h to finance, in addition to the optimal pay-off function $\widehat{X}_T(x)$, h units of the cash account, thus ending up with the pay-off function $\widehat{X}_T(x) + h$ at time T. Comparing this investment strategy to the optimal one corresponding to the initial endowment $x + h$, which is $\widehat{X}_T(x + h)$, we obtain

$$(2.57) \qquad \lim_{h\to 0}\frac{u(x+h) - u(x)}{h} = \lim_{h\to 0}\frac{\mathbf{E}[U(\widehat{X}_T(x+h)) - U(\widehat{X}_T(x))]}{h}$$

$$(2.58) \qquad \geq \lim_{h\to 0}\frac{\mathbf{E}[U(\widehat{X}_T(x) + h) - U(\widehat{X}_T(x))]}{h}$$

$$(2.59) \qquad = \mathbf{E}[U'(\widehat{X}_T(x))].$$

Using the fact that u is differentiable, and that h may be positive as well as negative, we must have equality in (2.58) and therefore have found another proof of formula (2.51) for $u'(x)$; the economic interpretation of this proof is that the economic agent, who is optimally investing, is indifferent of first order towards a (small) additional investment into the cash account.

Playing the same game as above, but using the additional endowment $h \in \mathbb{R}$ to finance an additional investment into the optimal portfolio $\widehat{X}_T(x)$ (assuming, for simplicity, $x \neq 0$), we arrive at the pay-off function $\frac{x+h}{x}\widehat{X}_T(x)$. Comparing this investment with $\widehat{X}_T(x + h)$, an analogous calculation as in (2.57) leads to the formula for $u'(x)$ displayed in (2.52). The interpretation now is, that the optimally investing economic agent is indifferent of first order towards a marginal variation of the investment into the optimal portfolio.

It now becomes clear that formulae (2.51) and (2.52) for $u'(x)$ are just special cases of a more general principle: for each $f \in L^\infty(\Omega, \mathcal{F}, \mathbf{P})$ we have

$$(2.60) \qquad \mathbf{E}_{\mathbf{Q}}[f]u'(x) = \lim_{h \to 0} \frac{\mathbf{E}_{\mathbf{P}}[U(\widehat{X}_T(x) + hf) - U(\widehat{X}_T(x))]}{h}.$$

The proof of this formula again is along the lines of (2.57) and the interpretation is the following: by investing an additional endowment $h\mathbf{E}_{\mathbf{Q}}[f]$ to finance the contingent claim hf, the increase in expected utility is of first order equal to $h\mathbf{E}_{\mathbf{Q}}[f]u'(x)$; hence again the economic agent is of first order indifferent towards an additional investment into the contingent claim f.

2.1.2. – The incomplete case

We now drop the assumption that the set $\mathcal{M}^e(S)$ of equivalent martingale measures is reduced to a singleton (but we still remain in the framework of a finite probability space Ω) and replace it by Assumption 1.1 requiring that $\mathcal{M}^e(S) \neq \emptyset$.

In this setting it follows from Theorem 2.11 that a random variable $X_T(\omega_n) = \xi_n$ may be dominated by a random variable of the form $x + (H \cdot S)_T$ iff $\mathbf{E}_{\mathbf{Q}}[X_T] = \sum_{n=1}^{N} q_n \xi_n \leq x$, for each $\mathbf{Q} = (q_1 \dots, q_N) \in \mathcal{M}^a(S)$ (or equivalently, for every $\mathbf{Q} \in \mathcal{M}^e(S)$).

In order to reduce the infinitely many constraints, where $\mathbf{Q}$ runs through $\mathcal{M}^a(S)$, to a finite number, make the easy observation that $\mathcal{M}^a(S)$ is a bounded, closed, convex polytope in $\mathbb{R}^N$ and therefore the convex hull of its finitely many extreme points $\{\mathbf{Q}^1, \dots, \mathbf{Q}^M\}$. Indeed, $\mathcal{M}^a(S)$ is given by finitely many linear constraints. For $1 \leq m \leq M$, we identify $\mathbf{Q}^m$ with the probabilites $(q_1^m, \dots, q_N^m)$.

Fixing the initial endowment $x \in \text{dom}(U)$, we therefore may write the utility maximization problem (1.16) similarly as in (2.19) as a concave optimization problem over $\mathbb{R}^N$ with finitely many linear constraints:

$$(2.61) \qquad \mathbf{E_P}[U(X_T)] = \sum_{n=1}^{N} p_n U(\xi_n) \rightarrow \max!$$

$$(2.62) \qquad \mathbf{E_{Q^m}}[X_T] = \sum_{n=1}^{N} q_n^m \xi_n \leq x, \text{ for } m = 1, \ldots, M.$$

Writing again

$$(2.63) \qquad C(x) = \{X_T \in L^0(\Omega, \mathcal{F}, \mathbf{P}) : \mathbf{E_Q}[X_T] \leq x, \text{ for all } \mathbf{Q} \in \mathcal{M}^a(S)\}$$

we define the value function, for $x \in \text{dom}(U)$,

$$(2.64) \qquad u(x) = \sup_{H \in \mathcal{H}} \mathbf{E}\big[U\big(x + (H \cdot S)_T\big)\big] = \sup_{X_T \in C(x)} \mathbf{E}[U(X_T)].$$

The Lagrangian now is given by

$$(2.65) \qquad L(\xi_1, \ldots, \xi_N, \eta_1, \ldots, \eta_M)$$

$$(2.66) \qquad = \sum_{n=1}^{N} p_n U(\xi_n) - \sum_{m=1}^{M} \eta_m \left(\sum_{n=1}^{N} q_n^m \xi_n - x \right)$$

$$(2.67) \qquad = \sum_{n=1}^{N} p_n \left(U(\xi_n) - \sum_{m=1}^{M} \frac{\eta_m q_n^m}{p_n} \xi_n \right) + \sum_{m=1}^{M} \eta_m x,$$

$$(2.68) \qquad \text{where } (\xi_1, \ldots, \xi_N) \in \text{dom}(U)^N, \quad (\eta_1, \ldots, \eta_M) \in \mathbb{R}_+^M.$$

Writing $y = \eta_1 + \ldots + \eta_M$, $\mu_m = \frac{\eta_m}{y}$, $\mu = (\mu_1, \ldots, \mu_m)$ and

$$(2.69) \qquad \mathbf{Q}^\mu = \sum_{m=1}^{M} \mu_m \mathbf{Q}^m,$$

note that, when $(\eta_1, \ldots, \eta_M)$ runs trough $\mathbb{R}_+^M$, the pairs $(y, \mathbf{Q}^\mu)$ run through $\mathbb{R}_+ \times \mathcal{M}^a(S)$. Hence we may write the Lagrangian as

$$L(\xi_1, \ldots, \xi_N, y, \mathbf{Q}) = \mathbf{E_P}[U(X_T)] - y\big(\mathbf{E_Q}[X_T - x]\big)$$

$$(2.70) \qquad = \sum_{n=1}^{N} p_n \left(U(\xi_n) - \frac{y q_n}{p_n} \xi_n \right) + yx,$$

where $\xi_n \in \text{dom}(U)$, $y > 0$, $\mathbf{Q} = (q_1, \ldots, q_N) \in \mathcal{M}^a(S)$.

This expression is entirely analogous to (2.22), the only difference now being that $\mathbf{Q}$ runs through the set $\mathcal{M}^a(S)$ instead of being a fixed probability measure. Defining again

$$(2.71) \qquad \Phi(\xi_1, \dots, \xi_n) = \inf_{y>0, \mathbf{Q} \in \mathcal{M}^a(S)} L(\xi_1, \dots, \xi_N, y, \mathbf{Q}),$$

and

$$(2.72) \qquad \Psi(y, \mathbf{Q}) = \sup_{\xi_1, \dots, \xi_N} L(\xi_1, \dots, \xi_N, y, \mathbf{Q}),$$

we obtain, just as in the complete case,

$$(2.73) \qquad \sup_{\xi_1, \dots, \xi_N} \Phi(\xi_1, \dots, \xi_N) = u(x), \quad x \in \mathrm{dom}(U),$$

and

$$(2.74) \qquad \Psi(y, \mathbf{Q}) = \sum_{n=1}^{N} p_n V\left(\frac{y q_n}{p_n}\right) + yx, \quad y > 0, \quad \mathbf{Q} \in \mathcal{M}^a(S),$$

where $(q_1, \dots, q_N)$ denotes the probabilities of $\mathbf{Q} \in \mathcal{M}^a(S)$. The minimization of Ψ will be done in two steps: first we fix $y > 0$ and minimize over $\mathcal{M}^a(S)$, i.e.,

$$(2.75) \qquad \Psi(y) := \inf_{\mathbf{Q} \in \mathcal{M}^a(S)} \Psi(y, \mathbf{Q}), \quad y > 0.$$

For fixed $y > 0$, the continuous function $\mathbf{Q} \to \Psi(y, \mathbf{Q})$ attains its minimum on the compact set $\mathcal{M}^a(S)$, and the minimizer $\widehat{\mathbf{Q}}(y)$ is unique by the strict convexity of V. Writing $\widehat{\mathbf{Q}}(y) = (\widehat{q}_1(y), \dots, \widehat{q}_N(y))$ for the minimizer, it follows from $V'(0) = -\infty$ that $\widehat{q}_n(y) > 0$, for each $n = 1, \dots, N$; Indeed, suppose that $\widehat{q}_n(y) = 0$, for some $1 \le n \le N$ and fix any equivalent martingale measure $\mathbf{Q} \in \mathcal{M}^e(S)$. Letting $\mathbf{Q}^\epsilon = \epsilon \mathbf{Q} + (1 - \epsilon)\widehat{\mathbf{Q}}$ we have that $\mathbf{Q}^\epsilon \in \mathcal{M}^e(S)$, for $0 < \epsilon < 1$, and $\Psi(y, \mathbf{Q}^\epsilon) < \Psi(y, \widehat{\mathbf{Q}})$ for $\epsilon > 0$ sufficiently small, a contradiction. In other words, $\widehat{\mathbf{Q}}(y)$ is an equivalent martingale measure for S.

Defining the dual value function $v(y)$ by

$$(2.76) \qquad v(y) = \inf_{\mathbf{Q} \in \mathcal{M}^a(S)} \sum_{n=1}^{N} p_n V\left(y \frac{q_n}{p_n}\right)$$

$$(2.77) \qquad = \sum_{n=1}^{N} p_n V\left(y \frac{\widehat{q}_n(y)}{p_n}\right)$$

we find ourselves in an analogous situation as in the complete case above: defining again $\widehat{y}(x)$ by $v'(\widehat{y}(x)) = -x$ and

$$(2.78) \qquad \widehat{\xi}_n = I\left(\widehat{y}(x)\frac{\widehat{q}_n(y)}{p_n}\right),$$

similar arguments as above apply to show that $(\widehat{\xi}_1, \dots, \widehat{\xi}_N, \widehat{y}(x), \widehat{\mathbf{Q}}(y))$ is the unique saddle-point of the Lagrangian (2.70) and that the value functions u and v are conjugate.

Let us summarize what we have found in the incomplete case:

THEOREM 2.18 (finite Ω, incomplete market). *Let the financial market* $S = (S_t)_{t=0}^T$ *defined over the finite filtered probability space* $(\Omega, \mathcal{F}, (\mathcal{F})_{t=0}^T, \mathbf{P})$ *and let* $\mathcal{M}^e(S) \neq \emptyset$, *and the utility function U satisfies Assumptions 1.2.*

Denote by $u(x)$ and $v(y)$ the value functions

$$(2.79) \qquad u(x) = \sup_{X_T \in C(x)} \mathbf{E}[U(X_T)], \qquad x \in \mathrm{dom}(U),$$

$$(2.80) \qquad v(y) = \inf_{Q \in \mathcal{M}^a(S)} \mathbf{E}\left[V\left(y\frac{d\mathbf{Q}}{d\mathbf{P}}\right)\right], \quad y > 0.$$

We then have:

(i) *The value functions $u(x)$ and $v(y)$ are conjugate and u shares the qualitative properties of U listed in Assumption 1.2.*

(ii) *The optimizers $\widehat{X}_T(x)$ and $\widehat{\mathbf{Q}}(y)$ in (2.79) and (2.80) exist, are unique, $\widehat{\mathbf{Q}}(y) \in \mathcal{M}^e(S)$, and satisfy*

$$(2.81) \qquad \widehat{X}_T(x) = I\left(y\frac{d\widehat{\mathbf{Q}}(y)}{d\mathbf{P}}\right), \quad y\frac{d\widehat{\mathbf{Q}}(y)}{d\mathbf{P}} = U'(\widehat{X}_T(x)),$$

where $x \in \mathrm{dom}(U)$ and $y > 0$ are related via $u'(x) = y$ or, equivalently, $x = -v'(y)$.

(iii) *The following formulae for u' and v' hold true*:

$$(2.82) \qquad u'(x) = \mathbf{E}_\mathbf{P}[U'(\widehat{X}_T(x))], \quad v'(y) = \mathbf{E}_{\widehat{Q}}\left[V'\left(y\frac{d\widehat{\mathbf{Q}}(y)}{d\mathbf{P}}\right)\right]$$

$$(2.83) \qquad \begin{aligned} xu'(x) &= \mathbf{E}_\mathbf{P}[\widehat{X}_T(x)U'(\widehat{X}_T(x))], \\ yv'(y) &= \mathbf{E}_\mathbf{P}\left[y\frac{d\widehat{\mathbf{Q}}(y)}{d\mathbf{P}}V'\left(y\frac{d\widehat{\mathbf{Q}}(y)}{d\mathbf{P}}\right)\right]. \end{aligned}$$

REMARK 2.19. Let us again interpret the formulae (2.82), (2.83) for $u'(x)$ similarly as in Remark 2.17 above. In fact, the interpretations of these formulae as well as their derivations remain in the incomplete case exactly the same.

But a new and interesting phenomenon arises when we pass to the variation of the optimal pay-off function $\widehat{X}_T(x)$ by a small unit of an arbitrary pay-off function $f \in L^\infty(\Omega, \mathcal{F}, \mathbf{P})$. Similarly as in (2.60) we have the formula

$$(2.84) \qquad \mathbf{E}_{\widehat{\mathbf{Q}}(y)}[f]u'(x) = \lim_{h \to 0} \frac{\mathbf{E}_\mathbf{P}[U(\widehat{X}_T(x) + hf) - U(\widehat{X}_T(x))]}{h},$$

the only difference being that $\mathbf{Q}$ has been replaced by $\widehat{\mathbf{Q}}(y)$ (recall that x and y are related via $u'(x) = y$).

The remarkable feature of this formula is that it does not only pertain to variations of the form $f = x + (H \cdot S)_T$, i.e, contingent claims attainable at price x, but to arbitrary contingent claims f, for which – in general – we cannot derive the price from no arbitrage considerations.

The economic interpretation of formula (2.84) is the following: the pricing rule $f \mapsto \mathbf{E}_{\widehat{\mathbf{Q}}(y)}[f]$ yields precisely those prices, at which an economic agent with initial endowment x, utility function U and investing optimally, is indifferent of first order towards adding a (small) unit of the contingent claim f to her portfolio $\widehat{X}_T(x)$.

In fact, one may turn the view around, and this was done by M. Davis [D97] (compare also the work of L. Foldes [F90]): one may *define* $\widehat{\mathbf{Q}}(y)$ by (2.84), verify that this indeed is an equivalent martingale measure for S, and interpret this pricing rule as "pricing by marginal utility", which is, of course, a classical and basic paradigm in economics.

Let us give a proof for (2.84) (under the hypotheses of Theorem 2.18). One possible strategy of proof, which also has the advantage of a nice economic interpretation, is the idea of introducing "fictitious securities" as developed in [KLSX91]: fix $x \in \text{dom}(U)$ and $y = u'(x)$ and let $(f^1, \dots, f^k)$ be finitely many elements of $L^\infty(\Omega, \mathcal{F}, \mathbf{P})$ such that the space $K = \{(H \cdot S)_T : H \in \mathcal{H}\}$, the constant function $\mathbf{1}$, and $(f^1, \dots, f^k)$ linearly span $L^\infty(\Omega, \mathcal{F}, \mathbf{P})$. Define the k processes

$$(2.85) \qquad S_t^{d+j} = \mathbf{E}_{\widehat{\mathbf{Q}}(y)}[f^j | \mathcal{F}_t], \quad j = 1, \dots, k, \quad t = 0, \dots, T.$$

Now extend the $\mathbb{R}^{d+1}$-valued process $S = (S^0, S^1, \dots, S^d)$ to the $\mathbb{R}^{d+k+1}$-valued process $\overline{S} = (S^0, S^1, \dots, S^d, S^{d+1}, \dots, S^{d+k})$ by adding these new coordinates. By (2.85) we still have that $\overline{S}$ is a martingale under $\widehat{\mathbf{Q}}(y)$, which now is the unique probability under which $\overline{S}$ is a martingale, by our choice of $(f^1, \dots, f^k)$ and Corollary 2.12.

Hence we find ourselves in the situation of Theorem 2.16. By comparing (2.50) and (2.81) we observe that the optimal pay-off function $\widehat{X}_T(x)$ has not changed. Economically speaking this means that in the "completed" market $\overline{S}$ the optimal investment may still be achieved by trading only in the first $d+1$ assets and without touching the "fictitious" securities $S^{d+1}, \dots, S^{d+k}$.

In particular, we now may apply formula (2.60) to $\mathbf{Q} = \widehat{\mathbf{Q}}(y)$ to obtain (2.84).

Finally we remark that the pricing rule induced by $\widehat{\mathbf{Q}}(y)$ is precisely such that the interpretation of the optimal investment $\widehat{X}_T(x)$ defined in (2.81) (given in Remark 2.17 in terms of marginal utility and the ratio of Arrow prices $\widehat{q}_n(y)$ and probabilities p_n) carries over to the present incomplete setting. The above completion of the market by introducing "fictious securities" allows for an economic interpretation of this fact.

CHAPTER III

The General Case

In the previous chapter we have analyzed the duality theory of the utility maximization problem in detail and with full proofs, for the case when the underlying probability space is finite.

We now pass to the question under which conditions the crucial features of the above Theorem 2.18 carry over to the general setting. In particular one is naturally led to ask: under which conditions

- are the optimizers $\widehat{X}_T(x)$ and $\widehat{\mathbf{Q}}(y)$ of the value functions $u(x)$ and $v(y)$ attained?
- does the basic duality formula

$$(3.1) \qquad U'\big(\widehat{X}_T(x)\big) = \widehat{y}(x)\frac{d\widehat{\mathbf{Q}}(\widehat{y}(x))}{d\mathbf{P}}$$

or, equivalently

$$(3.2) \qquad \widehat{X}_T(x) = I\left(\widehat{y}(x)\frac{d\widehat{\mathbf{Q}}(\widehat{y}(x))}{d\mathbf{P}}\right)$$

hold true?

- are the value functions $u(x)$ and $v(y)$ conjugate?
- does the value function $u(x)$ still inherit the qualitative properties of U listed in Assumption 1.2?
- do the formulae for $u'(x)$ and $v'(y)$ still hold true?

We shall see that we get affirmative answers to these questions under two provisos: firstly, one has to make an appropriate choice of the sets in which X_T and $\mathbf{Q}$ are allowed to vary. This choice will be different for Case 1, where $\mathrm{dom}(U) = \mathbb{R}_+$, and Case 2, where $\mathrm{dom}(U) = \mathbb{R}$. Secondly, the utility function U has to satisfy – in addition to Assumption 1.2 – a mild regularity condition, namely the property of "*reasonable asymptotic elasticity*".

3.1. – The reasonable asymptotic elasticity condition

The essential message of the theorems below is that, assuming that U has *"reasonable asymptotic elasticity"*, the duality theory works just as well as in the case of finite Ω. On the other hand, we shall see that we do not have to impose any regularity conditions on the underlying stochastic process S, except for its arbitrage-freeness in the sense made precise by Assumption 1.1. We shall also see that the assumption of reasonable asymptotic elasticity on the utility function U cannot be relaxed, even if we impose very strong assumptions on the process S (e.g., having continuous paths and defining a complete financial market), as we shall see below.

Before passing to the positive results we first analyze the notion of "reasonable asymptotic elasticity" and sketch the announced counterexample.

DEFINITION 3.1. A utility function U satisfying Assumption 1.2 is said to have *"reasonable asymptotic elasticity"* if

$$(3.3) \qquad AE_{+\infty}(U) = \limsup_{x \to \infty} \frac{xU'(x)}{U(x)} < 1,$$

and, in Case 2 of Assumption 1.2, we also have

$$(3.4) \qquad AE_{-\infty}(U) = \liminf_{x \to -\infty} \frac{xU'(x)}{U(x)} > 1.$$

We recall the following lemma from [KS99, Lemma 6.1], from which it follows that, for any concave function U such that the right hand side makes sense, we always have that $AE_{+\infty}(U) \leq 1$. Note that, the asymptotic elasticity assumption requires that the strict inequality holds.

LEMMA 3.2. *For a strictly concave, increasing, real-valued differentiable function U the asymptotic elasticity $AE(U)$ is well-defined and, depending on $U(\infty) = \lim_{x \to \infty} U(x)$, takes its values in the following sets:*
  (i) *For $U(\infty) = \infty$ we have $AE(U) \in [0, 1]$,*
  (ii) *For $0 < U(\infty) < \infty$ we have $AE(U) = 0$,*
  (iii) *For $-\infty < U(\infty) \leq 0$ we have $AE(U) \in [-\infty, 0]$.*

PROOF. (i) Using the monotonicity and positivity of U' we may estimate

$$0 \leq xU'(x) = (x-1)U'(x) + U'(x)$$
$$\leq [U(x) - U(1)] + U'(1)$$

hence, in the case $U(\infty) = \infty$,

$$0 \leq \limsup_{x \to \infty} \frac{xU'(x)}{U(x)} \leq \limsup_{x \to \infty} \frac{U(x) - U(1) + U'(1)}{U(x)} = 1.$$

(ii) For each $x_0 > 0$ we have

$$\limsup_{x \to \infty} x U'(x) = \limsup_{x \to \infty} (x - x_0) U'(x)$$

$$\leq \limsup_{x \to \infty} (U(x) - U(x_0)).$$

If $U(\infty) < \infty$ we may choose x_0 such that the right hand side becomes arbitrary small.

(iii) We infer from $U(\infty) \leq 0$ that $U(x) < 0$, for $x \in \mathbb{R}_+$, so that $\frac{x U'(x)}{U(x)} < 0$, for all $x \in \mathbb{R}_+$. $\qquad \square$

EXAMPLE 3.3.
- For $U(x) = \log x$, we have $\mathrm{AE}_{+\infty}(U) = 0$.
- For $U(x) = \frac{x^\alpha}{\alpha}$, we have $\mathrm{AE}_{+\infty}(U) = \alpha$, for $\alpha \in (-\infty, 1) \setminus \{0\}$.
- For $U(x) = \frac{x}{\log x}$ for $x \geq x_0$, we have $\mathrm{AE}_{+\infty}(U) = 1$.

The asymptotic elasticity compares as follows with other conditions used in the literature [KLSX91]:

LEMMA 3.4. *Let U be a utility function, and consider the following conditions:*
i) *There exists $x_0 > 0$, $\alpha < 1$, $\beta > 1$ such that $U'(\beta x) < \alpha U'(x)$ for all $x \geq x_0$.*
ii) $\mathrm{AE}_{+\infty}(U) < 1$
iii) *There exist k_1, k_2 and $\gamma < 1$ such that $U(x) \leq k_1 + k_2 x^\gamma$ for all $x \geq 0$.*
Then we have that i) $\Rightarrow$ ii) $\Rightarrow$ iii). *The reverse implications do not hold true in general, but if $\lim_{x \to \infty} \frac{x U'(x)}{U(x)}$ exists, then* ii) $\Longleftrightarrow$ iii). *If $\lim_{x \to \infty} \frac{x U'(x)}{U(x)}$ exists and is strictly positive, then* i) $\Longleftrightarrow$ ii).

PROOF. (i) $\Rightarrow$ (ii). Assume (i) and let $a = \alpha\beta$ and $b = \frac{1}{\alpha} > 1$ and estimate, for $x > ax_0$:

$$(3.5) \qquad U(bx) = U(\beta x_0) + \int_{\beta x_0}^{bx} U'(t) dt$$

$$(3.6) \qquad = U(\beta x_0) + \beta \int_{x_0}^{x/a} U'(\beta t) dt$$

$$(3.7) \qquad \leq U(\beta x_0) + \alpha\beta \int_{x_0}^{x/a} U'(t) dt$$

$$(3.8) \qquad = U(\beta x_0) + aU\left(\frac{x}{a}\right) - aU(x_0).$$

It follows that criterion (ii) of Corollary 6.1 in [KS99] is satisfied, hence $\mathrm{AE}(U) < 1$.

(ii) $\Rightarrow$ (iii) is immediate from assertion (i) of Lemma 6.3 in [KS99].

(ii) $\not\Rightarrow$ (i): for $n \in \mathbb{N}$, let $x_n = 2^{2^n}$ and define the function $U(x)$ by letting $U(x_n) = 1 - \frac{1}{n}$ and to be linear on the intervals $[x_{n-1}, x_n]$; (for $0 < x \le x_1$ continue $U(x)$ in an arbitrary way, so that U satisfies (2.4)).

Clearly $U(x)$ fails (i) as for any $\beta > 1$ there are arbitrary large $x \in \mathbb{R}$ with $U'(\beta x) = U'(x)$. On the other hand, we have $U(\infty) = 1$ so that $AE(U) = 0$ by Lemma 3.2. Finally, note that in this counterexample the limit $\lim_{x\to\infty} \frac{xU'(x)}{U(x)}$ exists and equals zero.

The attentive reader might object that $U(x)$ is neither strictly concave nor differentiable. But it is obvious that one can slightly change the function to "smooth out" the kinks and to "strictly concavify" the straight lines so that the above conclusion still holds true.

(iii) $\not\Rightarrow$ (ii): let again $x_n = 2^{2^n}$ and consider the utility function $\tilde{U}(x) = x^{1/2}$. Define $U(x)$ by letting $U(x_n) = \tilde{U}(x_n)$, for $n = 0, 1, 2\ldots$ and to be linear on the intervals $[x_n, x_{n+1}]$; (for $0 < x \le x_1$ again continue $U(x)$ in an arbitrary way, so that U satisfies (2.4)).

Clearly $U(x)$ satisfies condition (iii) as U is dominated by $\tilde{U}(x) = x^{1/2}$.

To show that $AE(U) = 1$ let $x \in]x_{n-1}, x_n[$ and calculate the marginal utility U' at x:

$$U'(x) = \frac{U(x_n) - U(x_{n-1})}{x_n - x_{n-1}} = \frac{2^{2^{n-1}} - 2^{2^{n-2}}}{2^{2^n} - 2^{2^{n-1}}} = \frac{2^{2^{n-1}}(1 - 2^{-2^{n-2}})}{2^{2^n}(1 - 2^{-2^{n-1}})}$$
$$= 2^{-2^{n-1}}(1 + o(1)).$$

On the other hand we calculate the average utility at $x = x_n$:

$$\frac{U(x_n)}{x_n} = \frac{2^{2^{n-1}}}{2^{2^n}} = 2^{-2^{n-1}}.$$

Hence

$$AE_{+\infty}(U) = \limsup_{x\to\infty} \frac{xU'(x)}{U(x)} = 1.$$

As regards the lack of smoothness and strict concavity of U a similar remark applies as in (ii) $\not\Rightarrow$ (i) above.

(ii) $\Rightarrow$ (i) under the assumption that $\lim_{x\to\infty} \frac{xU'(x)}{U(x)} = \gamma > 0$. By Corollary 6.1 (ii) in [KS99], condition (ii) is equivalent to the following: the exists some $x_0 > 0$, $\lambda > 1$ and $c < 1$ such that

(3.9) $$U(\lambda x) < c\lambda U(x)$$

for all $x > x_0$. Since $\lim_{x\to\infty} \frac{xU'(x)}{U(x)} = \gamma$ for any $\varepsilon > 0$ there exists some other x_0' such that

(3.10) $$\frac{xU'(x)}{\gamma(1+\varepsilon)} \le U(x) \le \frac{xU'(x)}{\gamma(1-\varepsilon)}$$

for all $x > x_0'$. Relabeling $x_0 = \max(x_0, x_0')$, and substituting (3.10) into (3.9), we obtain that:

$$U'(\lambda x) \leq c\frac{1+\varepsilon}{1-\varepsilon}U'(x)$$

which is clearly equivalent to (i) by choosing ε small enough, so that $c\frac{1+\varepsilon}{1-\varepsilon} < 1$. Note that in the case $\gamma = 0$ the above argument does not hold true, as shown in the example (ii) $\not\Rightarrow$ (i) above.

(iii) $\Rightarrow$ (ii) under the assumption that $\lim_{x\to\infty}\frac{xU'(x)}{U(x)}$ exists. By contradiction, suppose that $AE_{+\infty}(U) = 1$. Then for all $\varepsilon > 0$ there exists some x_0 such that $\frac{xU'(x)}{U(x)} > 1 - \varepsilon$ for all $x > x_0$. It follows that:

$$\log U(x) - \log U(x_0) = \int_{x_0}^{x}(\log U)'(y)dy \geq (1-\varepsilon)\int_{x_0}^{x}\frac{dy}{y}$$
$$= (1-\varepsilon)(\log x - \log x_0)$$

and hence:

(3.11) $$U(x) > U(x_0)\left(\frac{x}{x_0}\right)^{1-\varepsilon}.$$

Choosing $\varepsilon < 1 - \gamma$, as $x \to \infty$ (3.11) clearly contradicts (iii). $\quad\square$

Let us discuss the economic meaning of the notion of reasonable asymptotic elasticity: as H.-U. Gerber ponted out to us, the quantity $\frac{xU'(x)}{U(x)}$ is the elasticity of the function U at x. We are interested in its asymptotic behaviour. It easily follows from Assumption 1.2 that the limits in (3.3) and (3.4) are less (resp. bigger) than or equal to one (compare Lemma 3.2). What does it mean that $\frac{xU'(x)}{U(x)}$ tends to one, for $x \to \infty$? It means that the ratio between the *marginal utility* $U'(x)$ and the *average utility* $\frac{U(x)}{x}$ tends to one. A typical example is a function $U(x)$ which equals $\frac{x}{\ln(x)}$, for x large enough; note however, that in this example Assumption 1.2 is not violated insofar as the marginal utility still decreases to zero for $x \to \infty$, i.e., $\lim_{x\to\infty} U'(x) = 0$.

If the marginal utility $U'(x)$ is approximately equal to the average utility $\frac{U(x)}{x}$ for large x, this means that for an economic agent, modeled by the utility function U, the increase in utility by varying wealth from x to $x+1$, when x is large, is approximately equal to the average of the increase of utility by changing wealth from n to $n+1$, where n runs through $1, 2, \ldots, x-1$ (we assume in this argument that x is a large natural number and, w.l.o.g., that $U(1) \approx 0$). We feel that the economic intuition behind decreasing marginal utility suggests that, for large x, the marginal utility $U'(x)$ should be substantially smaller than the average utility $\frac{U(x)}{x}$. Therefore we have denoted a utility function, where the ratio of $U'(x)$ and $\frac{U(x)}{x}$ becomes arbitrarily close to one if x tends either to $+\infty$ or $-\infty$, as being "unreasonable". Another justification for this terminology will be the results of Theorems 3.15 and 3.19 below.

P. Guasoni observed, that there is a close connection between the asymptotic behaviour of the *elasticity* of U, and the asymptotic behaviour of the *relative risk aversion* associated to U. Recall (see, e.g., [HL88]) that the relative risk aversion of an agent with endowment x, whose preferences are described by the utility function U, equals

$$(3.12) \qquad\qquad RRA(U)(x) = -\frac{xU''(x)}{U'(x)}.$$

A formal application of de l'Hôpital's rule yields

$$(3.13) \qquad \lim_{x\to\infty} \frac{xU'(x)}{U(x)} = \lim_{x\to\infty} \frac{U'(x) + xU''(x)}{U'(x)} = 1 - \lim_{x\to\infty} \left(-\frac{xU''(x)}{U'(x)} \right)$$

which insinuates that the asymptotic elasticity of U is less than one iff the "*asymptotic relative risk aversion*" is strictly positive.

Turning the above formal argument into a precise statement, one easily proves the following result (Proposition B.1 below): if $\lim_{x\to\infty} (-\frac{xU''(x)}{U'(x)})$ exists, then $\lim_{x\to\infty} \frac{xU'(x)}{U(x)}$ exists too, and the former is strictly positive iff the latter is less than one. Hence "*essentially*" these two concepts coincide.

On the other hand, in general (i.e. without assuming that the above limit exists), there is no way to characterize the condition $\lim\sup_{x\to\infty} \frac{xU'(x)}{U(x)} < 1$ in terms of the asymptotic behaviour of $-\frac{xU''(x)}{U'(x)}$, as $x\to\infty$ (for more details on this issue, see appendix B).

Similar reasoning applies to the asymptotic behaviour of $\frac{xU'(x)}{U(x)}$, as x tends to $-\infty$, in Case 2. In this context the typical counter-example is $U(x) \sim x\ln(|x|)$, for $x < x_0$; in this case one finds similarly

$$(3.14) \qquad \lim_{x\to-\infty} U'(x) = \infty, \quad \text{while} \quad \lim_{x\to-\infty} \frac{xU'(x)}{U(x)} = 1.$$

The message of Definition 3.1 above is – roughly speaking – that we want to exclude utility functions U which behave like $U(x) \sim \frac{x}{\ln(x)}$, as $x\to\infty$, or $U(x) \sim x\ln|x|$, as $x\to-\infty$. Similar (but not quite equivalent) notions comparing the behaviour of $U(x)$ with that of power functions in the setting of Case 1, were defined and analyzed in [KLSX91] (see Lemma 3.4 above or [KS99], Lemma 6.5, for a comparison of these concepts).

3.2. – Counterexamples

We start with a counterexample showing the relevance of the notion of asymptotic elasticity in the context of utility maximization: *whenever U fails to have reasonable asymptotic elasticity the duality theory breaks down in a rather* dramatic way. We only state the version of the counterexample when the lim sup and the lim inf in Definition 3.1 are indeed limits and are bothe equal to one; we refer to [KS99] and [S03] for the general case (which only differs in some technicalities).

In fact for this counterexample we do not have to go very far in the degree of complexity of the price process S. We shall see that it suffices to take S to be a Geometric Brownian Motion *stopped at an appropriately chosen stopping time·* ([KS99], Example 5.2).

EXAMPLE 3.5 ([S01], Proposition 3.5). Let U be *any utility function* satisfying Assumption 1.2, Case 2 and such that

$$(3.15) \qquad \lim_{x \to -\infty} \frac{x U'(x)}{U(x)} = \lim_{x \to \infty} \frac{x U'(x)}{U(x)} = 1.$$

Then there is an $\mathbb{R}$-valued process $(S_t)_{0 \le t \le T}$ of the form

$$(3.16) \qquad S_t = \exp \left(B_t + \mu_t \right),$$

where $B = (B_t)_{0 \le t \le T}$ is a standard Brownian motion, based on its natural filtered probability space, and μ_t a predictable process, such that the following properties hold true:

  (i) $\mathcal{M}^e(S) = \{\mathbf{Q}\}$, i.e., S defines a complete financial market.
 (ii) The primal value function $u(x)$ fails to be strictly concave and to satisfy $u'(\infty) = 0$, $u'(-\infty) = \infty$ in a rather striking way: $u(x)$ is a straight line of the form $u(x) = c + x$, for some constant $c \in \mathbb{R}$.
(iii) The optimal investment $\widehat{X}_T(x)$ fails to exist, for all $x \in \mathbb{R}$, except for one point $x = x_0$. In particular, for $x \ne x_0$, the formula (3.2) does not define the optimal investment $\widehat{X}_T(x)$.
 (iv) The dual value function v fails to be a finite, smooth, strictly convex function on $\mathbb{R}_+$ in a rather striking way: in fact, $v(1) = c < \infty$ while $v(y) = \infty$, for all $y \ne 1$.

We shall try to sketch the basic idea underlying the construction of the example, in mathematical as well as economic terms. Arguing mathematically, one starts by translating the assumptions (3.15) on the utility function U into equivalent properties of the conjugate function V: roughly speaking, the corresponding property of $V(y)$ is, that it increases very rapidly to infinity, as $y \to 0$ and $y \to \infty$ (see [KS99, Corollary 6.1] and [S01, Proposition 4.1]). Having isolated this property of V, it is an easy exercise to construct a function $f : [0, 1] \to]0, \infty[$, $\mathbf{E}[f] = 1$ such that

$$(3.17) \qquad c : \mathbf{E}\big[V(f)\big] < \infty \quad \text{while} \quad \mathbf{E}\big[V(yf)\big] = \infty, \quad \text{for} \quad y \ne 1,$$

where $\mathbf{E}$ denotes expectation with respect to Lebesque measure λ. In fact one may find such a function f taking only the values $(y_n)_{n=-\infty}^{\infty}$, for a suitable chosen increasing sequence $(y_n)_{n=-\infty}^{\infty}$, $\lim_{n \to -\infty} y_n = 0$, $\lim_{n \to \infty} y_n = \infty$.

Next we construct a measure $\mathbf{Q}$ on the sigma algebra $\mathcal{F} = \mathcal{F}_T$ generated by the Brownian motion $B = (B_t)_{0 \leq t \leq T}$ which is equivalent to Wiener measure $\mathbf{P}$, and such that the distribution of $\frac{d\mathbf{Q}}{d\mathbf{P}}$ (under $\mathbf{P}$) equals that of f (under Lebesgue measure λ). There is no uniqueness in this part of the construction, but it is straightforward to find some appropriate measure $\mathbf{Q}$ with this property.

By Girsanov's theorem we know that we can find an adapted process $(\mu_t)_{0 \leq t \leq T}$, such that $\mathbf{Q}$ is the unique equivalent local martingale measure for the process defined in (3.16), hence we obtain assertion (i).

This construction makes sure that we obtain property (iv), i.e.

$$(3.18) \qquad v(y) = \mathbf{E}_\mathbf{P}\left[V\left(y\frac{d\mathbf{Q}}{d\mathbf{P}}\right)\right] = \mathbf{E}_\lambda[V(yf)] < \infty \quad \text{iff} \quad y = 1.$$

Once this crucial property is established, most of the assertions made in (ii) and (iii) above easily follow (in fact, for the existence of $\widehat{X}_T(x)$ for precisely one $x = x_0$, some extra care is needed): for esample, the function $u(x) = c + x$ is such that $v(y)$ is conjugate to u, which – at least formally – yields (ii).

Instead of elaborating further on the mathematical details of the construction sketched above, let us try to give an economic interpretation of what is really happening in the above example. This is not easy, but we find it worth trying. We concentrate on the behaviour of U as $x \to \infty$, the case when $x \to -\infty$ being similar.

How is the "unreasonability" property of the utility function U used to construct the pathologies in the above example? Here is a rough indication of the underlying economic idea: the financial market S is constructed in such a way that one may find positive numbers $(x_n)_{n=1}^\infty$, disjoint sets $(A_n)_{n=1}^\infty$ in $\mathcal{F}_T$ (which correspond to the sets $\{f = y_n\}$ in the above sketch), with $\mathbf{P}[A_n] = p_n$ and $\mathbf{Q}[A_n] = q_n$, such that for the contingent claims $x_n \mathbf{1}_{A_n}$ we approximately have

$$(3.19) \qquad \mathbf{E}_\mathbf{Q}[x_n \mathbf{1}_{A_n}] = q_n x_n \approx 1$$

and

$$(3.20) \qquad \mathbf{E}_\mathbf{P}[U(x_n)\mathbf{1}_{A_n}] = p_n U(x_n) \approx 1.$$

Hence $\frac{q_n}{p_n} \approx \frac{U(x_n)}{x_n}$.

It is easy to construct a complete, continuous market S (e.g., over the Brownian filtration) such that this situation occurs and this is, in fact, what is done in the above "mathematical" argument to define f and $\mathbf{Q}$.

We claim that, for any $x \in \mathbb{R}$ and any investment strategy $X_T = x + (H \cdot S)_T$, we can find an investment strategy $\widetilde{X}_T = (x + 1) + (\widetilde{H} \cdot S)_T$ such that

$$(3.21) \qquad \mathbf{E}[U(\widetilde{X}_T)] \approx \mathbf{E}[U(X_T)] + 1.$$

The above relation should motivate why the value function $u(x)$ becomes a straight line with slope one, at least for x sufficiently large (for the corresponding behaviour of $u(x)$ on the left hand side of $\mathbb{R}$ one has to play in addition a similar game as above with $(x_n)_{n=1}^{\infty}$ tending to $-\infty$).

To present the idea behind (3.21), suppose that we have $\mathbf{E}[U(X_T)] < \infty$, so that $\lim_{n \to \infty} \mathbf{E}[U(X_T)\mathbf{1}_{A_n}] = 0$. Varying our initial endowment from x to $x + 1$ Euro, we may use the additional Euro to add to the pay-off function X_T the function $x_n\mathbf{1}_{A_n}$, for some large n; by (3.19) this may be financed (approximately) with the additional Euro and by (3.20) this will increase the expected utility (approximately) by 1

$$
\begin{aligned}
\mathbf{E}\big[U(X_T + x_n\mathbf{1}_{A_n})\big] &\approx \mathbf{E}\big[U(X_T)\mathbf{1}_{\Omega \setminus A_n}\big] + \mathbf{E}\big[U(X_T + x_n)\mathbf{1}_{A_n}\big] \\
&\approx \mathbf{E}\big[U(X_T)\big] + p_n U(x_n) \\
&\approx \mathbf{E}\big[U(X_T)\big] + 1,
\end{aligned}
$$
(3.22)

which was claimed in (3.21).

The above argument also gives a hint why we cannot expect that the optimal strategy $\widehat{X}_T(x) = x + (\widehat{H}{\cdot}S)_T$ exists, as one cannot "pass to the limit as $n \to \infty$" in the above reasoning.

Observe that we have not yet used the assumption $\limsup_{x \to \infty} \frac{xU'(x)}{U(x)} = 1$, as it always is possible to construct things in such a way that (3.19) and (3.20) hold true. How does the "unreasonable asymptotic elasticity" come into play? The point is that we have to do the construction described in (3.19) and (3.20) without violating Assumption 1.3, i.e.,

$$
u(x) = \sup_{H \in \mathcal{H}} \mathbf{E}\big[U\big(x + (H \cdot S)_T\big)\big] < \infty,
$$
(3.23)
$$
\text{for some (equivalently, for all)} \quad x \in \mathbb{R}.
$$

In order to satisfy Assumption 1.3 we have to make sure – as a necessary condition – that

$$
\mathbf{E}\left[\sum_{n=1}^{\infty} U(\mu_n x_n)\mathbf{1}_{A_n}\right] = \sum_{n=1}^{\infty} p_n U(\mu_n x_n)
$$
(3.24)

remains bounded, when $(\mu_n)_{n=1}^{\infty}$ runs through all convex weights $\mu_n \geq 0$, $\sum_{n=1}^{\infty} \mu_n = 1$, i.e., when we consider all investments into non-negative linear combinations of the contingent claims $x_n\mathbf{1}_{A_n}$, which can be financed with one Euro.

The message of Example 3.5 is that this is possible, if and only if $\limsup_{x \to \infty} \frac{xU'(x)}{U(x)} = 1$ (for this part of the construction we only use the asymptotic behaviour of $U(x)$, as $x \to \infty$). To motivate this claim, think for a moment of the "reasonable" case, e.g., $U(x) = \frac{x^\alpha}{\alpha}$, for some $0 < \alpha < 1$, in which case we have $\lim_{x \to \infty} \frac{xU'(x)}{U(x)} = \alpha < 1$. Letting $\mu_n \approx c_\varepsilon n^{-(1+\epsilon)}$ (where the normalizing constant $c_\varepsilon > 0$ is chosen such that $\sum_{n=1}^\infty \mu_n = 1$), we get from (3.20)

$$(3.25) \qquad \sum_{n=1}^\infty p_n U(\mu_n x_n) \approx \sum_{n=1}^\infty n^{-(1+\epsilon)\alpha} p_n U(x_n)$$

$$(3.26) \qquad\qquad\qquad \approx \sum_{n=1}^\infty n^{-(1+\epsilon)\alpha},$$

which equals infinity if $\epsilon > 0$ is small enough, that $(1+\epsilon)\alpha \leq 1$. This argument indicates that in the case of the power utility $U(x) = \frac{x^\alpha}{\alpha}$ it is impossible to reconcile the validity of (3.19) and (3.20) with the requirement (3.23). On the other hand, it turns out that in the "unreasonable" case, where we have $\lim_{x\to\infty} \frac{xU'(x)}{U(x)} = 1$, we can do the construction in such a way that $U(\mu_n x_n)$ is sufficiently close to $\mu_n U(x_n)$ such that we obtain from (3.20) a uniform bound on the sum in (3.24).

After these motivating heuristic remarks, let us now pass to the formal construction of a counterexample similar in spirit to that of Example 3.5, but now modeled in countable discrete time.

It is convenient to introduce some notation.

DEFINITION 3.6. Let $t_n = 1 - \frac{1}{n+1}$. A financial market $(\Omega, \mathcal{F}, (\mathcal{F}_{t_n})_{n=0}^\infty, \mathbf{P}, (S_{t_n})_{n=0}^\infty)$ is a *simple jump model* if:

(i) $\Omega = \left(\cup_{n=1}^\infty A_n\right) \cup B$, where $(A_n)_{n=1}^\infty$ and B are disjoint nonempty sets such that $\mathbf{P}(A_n) = p_n$ for some numbers $p_n > 0$, where $\sum_{n=1}^\infty p_n = 1 - \mathbf{P}(B) < 1$.

(ii) $\mathcal{F}_{t_n} = \sigma((A_k)_{k=1}^n)$, for $n \geq 0$.

(iii) The risky asset process S_{t_n} is defined by:

$$S_{t_n} = \sum_{k=1}^n \alpha_k (\mathbf{1}_{A_k} - q_k) \mathbf{1}_{\Omega \setminus (\cup_{i=1}^{k-1} A_i)}$$

where the numbers $(\alpha_n)_{n=1}^\infty$ and $(q_n)_{n=1}^\infty$ satisfy $\sum_{n=1}^\infty \alpha_n < \infty$, $q_n > 0$ for all n, and $\sum_{n=1}^\infty q_n < 1$.

REMARK 3.7. It is immediate from the above definition that a *simple jump model* is complete, and the unique equivalent martingale measure $\mathbf{Q}$ for S is given by $\mathbf{Q}(A_n) = q_n$ and $\mathbf{Q}(B) = 1 - \sum_{n=1}^\infty q_n$.

We prove the following:

PROPOSITION 3.8. *Let U satisfy Assumption 1.2 and*

$$(3.27) \qquad \limsup_{x\to\infty} \frac{xU'(x)}{U(x)} = 1.$$

For any $x^* > 0$, there is a simple jump model $(S_{t_n})_{n=0}^{\infty}$, based on a countable filtered probability space $(\Omega, \mathcal{F}, (\mathcal{F}_{t_n})_{n=0}^{\infty}, \mathbf{P})$ such that:

(i) *$u(x) < \infty$, for all $x \in \text{dom}(U)$,*

(ii) *For $x > 0$, the utility maximization problem admits an optimizer $\widehat{X}(x) = x + (\widehat{H}(x) \cdot S)_T$, with $\widehat{H}(x)$ an admissible predictable process, if and only if $x \leq x^*$,*

(iii) *$u'(x)$ is constant for $x \geq x^*$.*

We break the proof of Proposition 3.8 into two lemmata. In the first one, we show that the quantities $(p_n)_{n=1}^{\infty}$ and $(q_n)_{n=1}^{\infty}$ in Definition 3.6 can be chosen so that the optimal terminal payoff on S^N (the process S stopped at the N-th step) with initial capital x^{**} prescribes an investment of $x^{**} - x_N$ in the Arrow-Debreu asset $\mathbf{1}_{A_N}$. As N increases to infinity, the value functions $u^N(x^{**})$ increase to $u(x^{**})$, also finite-valued, and y^N, the Lagrange multipliers associated to the above problems, increase to a finite value y^{∞}.

From now on, we shall write, for notational convenience, S_n (respectively H_n, X_n) in place of S_{t_n} (resp. H_{t_n}, X_{t_n}).

LEMMA 3.9. *Let U be a utility function satisfying Assumption 1.2 and (3.27). Let $x^* > 0$, $x^{**} > x^*$, and let $(x_n)_{n=1}^{\infty}$, with $x_1 > 0$, be a sequence strictly increasing to x^*. Consider a simple jump model, as in Definition 3.6, and denote by $(S_n^N)_{n=1}^{\infty}$ the process $(S_n)_{n=1}^{\infty}$ stopped at N, and by $u^N(x)$ and $u(x)$ the value functions of the utility maximization problems on S and S^N respectively. Finally, denote by $\widehat{X}_T^N(x^{**}) = \xi_1^N \mathbf{1}_{A_1} + \ldots + \xi_N^N \mathbf{1}_{A_N} + \mu^N \mathbf{1}_{B_N}$ the optimal terminal payoff on S^N with initial capital x^{**}.*
We can choose the quantities $(p_n)_{n=1}^{\infty}$ and $(q_n)_{n=1}^{\infty}$ such that:

i) $q_N \xi_N^N = x^{**} - x_N$

ii) $0 < q_n < \frac{p_n}{2}$ and $\sum_{n=1}^{\infty} p_n \leq \frac{1}{2}$.

iii) *$u(x) = \lim_{N \to \infty} u^N(x) < \infty$ for all $x > 0$, and $y^{\infty} = \lim_{N \to \infty} y^N < \infty$, where $y^N = (u^N)'(x^{**})$ and $y^{\infty} = u'(x^{**})$.*

PROOF. Note that without loss of generality we may assume $U(\infty) > 0$, as adding a constant to U does not change the optimization problem.

First we need to introduce some notation. Let us fix strictly positive numbers $(\varepsilon_n)_{n=1}^{\infty}$, such that $\prod_{n=1}^{\infty}(1 + \varepsilon_n) < 2$, and denote by $I = (U')^{-1}$. Denote also the set

$$\Xi = \left\{ x : U(x) > 0, \frac{xU'(x)}{U(x)} > \frac{1}{2} \right\}$$

which is open by Assumption 1.2 and unbounded by assumption (3.27).
We shall construct our model so that the following properties hold:

(a) $\sum_{n=1}^{N} p_n < \frac{1}{2} - 2^{-(N+1)}$

(b) $q_n < \frac{p_n}{2}$ for $1 \le n \le N$

(c) $\xi_N^N q_N = x^{**} - x_N$

(d) $U'\left(\frac{x_{N+1}-x_N}{2(x^{**}-x_N)}\xi_N^N\right) < U'(\xi_N^N)(1+\varepsilon_N)$

(e) $I(y\frac{q_N}{p_N}) \in \Xi$ for all $y \in [y^N, y^\infty]$.

We proceed by induction on N. Consider first $N = 1$: we shall find p_1 and q_1, which determine ξ_1^1, μ^1 and y^1 so that

$$\widehat{X}^1(x^{**}) = \xi_1^1 \mathbf{1}_{A_1} + \mu^1 \mathbf{1}_{B_1} \quad \text{and}$$
$$y^1 = (u^1)'(x^{**}).$$

For fixed parameter $\xi_1^1 \in \Xi$, we obtain the first order conditions for the optimal portfolio satisfying (c) for the one-period problem S^1 by solving the following equations for the unknowns p_1, q_1, y^1, μ^1, for given parameter ξ_1^1:

(3.28) $\xi_1^1 q_1 = x^{**} - x_1$

(3.29) $\xi_1^1 q_1 + \mu^1(1 - q_1) = x^{**}$

(3.30) $U'(\xi_1^1) = y^1 \frac{q_1}{p_1}$

(3.31) $U'(\mu^1) = y^1 \frac{1 - q_1}{1 - p_1}$

Let us check that this system has indeed a solution: q_1 is determined by (3.28), and μ^1 by (3.29). (3.30) determines p_1 in terms of y^1, and plugging in (3.31) we obtain:

(3.32) $$y^1 = \frac{U'(\mu^1)}{1 + \frac{x^{**} - x_1}{\xi_1^1}\left(\frac{1}{U'(\xi_1^1)} - 1\right)}.$$

Recalling that $\lim_{x\to\infty} \frac{U(x)}{x} = 0$ and that:

(3.33) $$\frac{1}{2} \le \liminf_{\substack{x\to\infty \\ x\in\Xi}} \frac{xU'(x)}{U(x)} \le \limsup_{\substack{x\to\infty \\ x\in\Xi}} \frac{xU'(x)}{U(x)} \le 1$$

we obtain that:

$$\lim_{\substack{\xi_1^1\to\infty \\ \xi_1^1\in\Xi}} y^1(\xi_1^1) = \lim_{\substack{\xi_1^1\to\infty \\ \xi_1^1\in\Xi}} \frac{U'(\mu^1)U(\xi_1^1)}{U(\xi_1^1) + (x^{**} - x_1)\left(\frac{U(\xi_1^1)}{\xi_1^1 U'(\xi_1^1)} - \frac{U(\xi_1^1)}{\xi_1^1}\right)} = \eta < \infty.$$

From (3.30) and (3.33) it follows that:

$$p_1 = y^1 \frac{x^{**} - x_1}{U'(\xi_1^1)\xi_1^1} < 2y^1 \frac{x^{**} - x_1}{U(\xi_1^1)}$$

and therefore we can assume that $p_1 < \frac{1}{4}$ and $q_1 < \frac{p_1}{2}$ (by (3.30)) for ξ_1^1 large enough.

Finally, we apply Lemma 3.11 to $\varepsilon = \min\left(\varepsilon_1, \frac{x_2 - x_1}{2(x^{**} - x_1)}\right)$, which provides some large ξ_1^1 for which (d) is satisfied, and condition (e) is obtained up to a change to smaller $(\varepsilon_n)_{n=2}^{\infty}$.

It is worthwhile to resume the present procedure. For the parameter ξ_1^1 we obtain from the necessary first-order conditions (3.28)-(3.31) how the quantities ξ_1^1, p_1, q_1, y^1 and μ^1 have to be related. Then we turn around and specify (for a large value of ξ_1^1) the parameters p_1, q_1 in our model as we have just obtained them. It then follows that for this model ξ_1^1 and μ^1 indeed define the optimal portfolio, while y^1 is the associated Lagrange multiplier satisfying $(u^1)'(x^{**}) = y^1$.

Let us now pass to the induction step. Suppose that we have constructed $(p_i)_{i=1}^N$ and $(q_i)_{i=1}^N$ such that the resulting quantities $(\xi_i^N)_{i=1}^N$ and $y^N = (u^N)'(x^{**})$ satisfy (a)–(e) above. Let us construct p_{N+1}, q_{N+1}, which in turn determine μ^{N+1}, y^{N+1} and $(\xi_i^{N+1})_{i=1}^{N+1}$ such that the same assumptions are satisfied for $N+1$.

Again, we leave $\xi = \xi_[^1 N + 1]$ as a free parameter to be fixed later, and solve the $N + 4$ equations:

(3.34) $$\xi_{N+1}^{N+1} N + 1 q_{N+1} = x^{**} - x_{N+1}$$

(3.35) $$\sum_{n=1}^{N} \xi_n^{N+1} q_n + \xi_[^1 N + 1] q_{N+1} + \mu^{N+1}\left(1 - \sum_{n=1}^{N+1} q_n\right) = x^{**}$$

(3.36) $$U'(\xi_n^{N+1}) = y^{N+1}\frac{q_n}{p_n} \quad \text{for} \quad n = 1 \ldots N + 1$$

(3.37) $$U'(\mu^{N+1}) = y^{N+1}\frac{1 - \sum_{n=1}^{N+1} q_n}{1 - \sum_{n=1}^{N+1} p_n}$$

in the unknowns q_{N+1}, p_{N+1}, y^{N+1}, μ^{N+1} and $(\xi_i^{N+1})_{i=1}^N$. Again, q_{N+1} is determined by (3.34). If we fix $y^{N+1} = y$ we are left with $N + 2$ equations ((3.36) and (3.37)), each one involving exactly one of the $N + 2$ unknowns $(\xi_i^{N+1})_{i=1}^N$, p_{N+1} and μ^{N+1}, which admit a unique solution depending on ξ and y. We write $\xi_n^{N+1}(y)$ and $\mu^{N+1}(y)$ to indicate their dependence on y. Now, denote the quantity on the left side of (3.35),

(3.38) $$G(y) = \sum_{n=1}^{N} \xi_n^{N+1}(y)q_n + (x^{**} - x_{N+1}) + \mu^{N+1}(y)\left(1 - \sum_{n=1}^{N+1} q_n\right).$$

Observe that G is a strictly decreasing function for $y \geq y^N$. In fact, from (3.36) to (3.37) we have that $\xi_n^{N+1} = I(y\frac{q_n}{p_n})$, and since I is a decreasing function, the first summation in (3.38) is decreasing. To prove the claim, it remains to show that $\mu^{N+1}(y)$ is also decreasing. Solving (3.36) for p_{N+1}, we have that:

$$p_{N+1} = y\frac{x^{**} - x_{N+1}}{\xi U'(\xi)}$$

and substituting in (3.37), we obtain:

$$\mu^{N+1}(y) = I\left(y\frac{1 - \sum_{n=1}^{N} q_n - \frac{x^{**} - x_{N+1}}{\xi}}{1 - \sum_{n=1}^{N} p_n - y\frac{x^{**} - x_{N+1}}{\xi U'(\xi)}}\right) = I\left(\frac{ay}{b - cy}\right)$$

where $a = 1 - \sum_{n=1}^{N} q_n - \frac{x^{**} - x_{N+1}}{\xi}$, $b = 1 - \sum_{n=1}^{N} p_n$, and $c = \frac{x^{**} - x_{N+1}}{\xi U'(\xi)}$ are all positive numbers. Since $\frac{ay}{b - cy}$ is an increasing function, the claim follows.

For $y = y^N$ we have that $\xi_n^{N+1}(y^N) = \xi_n^N$ for $n = 1, \dots, N$, and $\mu^{N+1}(y^N)$ and μ^N are arbitrarily close for large ξ. In fact, by the inductive hypothesis:

$$\sum_{n=1}^{N} \xi_n^N q_n + \mu^N\left(1 - \sum_{n=1}^{N} q_n\right) = x^{**}$$

we obtain that:

$$G(y^N) = x^{**} + (x^{**} - x_{N+1}) + \mu^{N+1}\left(1 - \sum_{n=1}^{N+1} q_n\right) - \mu^N\left(1 - \sum_{n=1}^{N} q_n\right).$$

For ξ large enough, μ^{N+1} approaches to μ^N and q_{N+1} approaches to zero, therefore we can assume that $G(y^N) > x^{**}$. For $y = y^N(1 + \varepsilon_n)$, since $\xi_n^{N+1}(y)$ is decreasing, we have that:

(3.39) $\xi_n^{N+1}(y^N(1 + \varepsilon_N)) \leq \xi_n^{N+1}(y^N) = \xi_n^N(y^N)$ for $n = 1 \dots N - 1$.

Furthermore, by the inductive hypothesis (d):

(3.40) $y^N(1 + \varepsilon_N) = U'(\xi_N^N)(1 + \varepsilon_N) > U'\left(\frac{x_{N+1} - x_N}{2(x^{**} - x_N)}\xi_N^N\right)$

and hence:

$$\xi_N^{N+1}(y(1 + \varepsilon_N)) < \frac{x_{N+1} - x_N}{2(x^{**} - x_N)}\xi_N^N.$$

Substituting (3.39) and (3.40) in (3.38), we obtain:

$$G(y^N(1+\varepsilon_N))$$

$$< \sum_{n=1}^{N-1} \xi_n^N q_n + x^{**} + \frac{x_{N+1} - x_N}{2(x^{**} - x_N)} \xi_N^N q_N + (x^{**} - x_{N+1})$$

$$+ \mu^{N+1}(y^N(1+\varepsilon_N)) \left(1 - \sum_{n=1}^{N+1} q_n\right)$$

$$= -(x^{**} - x_N) + \frac{x_{N+1} - x_N}{2} + (x^{**} - x_{N+1})$$

$$+ \mu^{N+1} \left(1 - \sum_{n=1}^{N+1} q_n\right) - \mu^N \left(1 - \sum_{n=1}^{N} q_n\right)$$

$$= -\frac{1}{2}(x_{N+1} - x_N) + \varphi(\xi)$$

where

$$\varphi(\xi) = \mu^{N+1}(\xi) \left(1 - \sum_{n=1}^{N} q_n - q_{N+1}(\xi)\right) - \mu^N \left(1 - \sum_{n=1}^{N} q_n\right)$$

is infinitesimal, in the sense that $\lim_{\xi \to \infty} \varphi(\xi) = 0$.

Therefore, for a sufficiently large ξ, $G(y^N(1+\varepsilon_N)) < x^{**}$, and there is a unique value $y^{N+1} \in (y^N, y^N(1+\varepsilon^N))$ such that $G(y) = x^{**}$. Hence, using this value of y^{N+1} in equations (3.34), (3.35), (3.36), (3.37), we obtain a solution to the entire system, and in particular we obtain the desired values for p_{N+1} and q_{N+1}. To check (b), (d) and (e), we apply Lemma 3.11 to $\varepsilon = \min\left(\varepsilon_N, \frac{x_{N+2} - x_{N+1}}{2(x^{**} - x_{N+1})}\right)$, and obtain some large ξ such that:

$$(3.41) \qquad U'\left(\frac{x_{N+2} - x_{N+1}}{2(x^{**} - x_{N+1})}\xi\right) < U'(\xi)(1 + \varepsilon_{N+1})$$

which implies (d), and up a change to smaller $(\varepsilon_n)_{n=N+1}^{\infty}$ we obtain (e). Finally, we let ξ be large enough so that $q_{N+1} < \frac{p_{N+1}}{2}$, and the induction hypotheses are satisfied. This completes the construction.

The sequence $(y^N)_{N=1}^{\infty}$ increases to $y^{\infty} < \infty$. In fact, the assumption $\prod_{n=1}^{\infty}(1+\varepsilon_n) < 2$ guarantees that:

$$y^N \leq y^1 \prod_{n=2}^{N}(1+\varepsilon_n) \leq 2y^1.$$

Then by (3.36) it follows that $(\xi_n^N)_{N=n}^\infty$ decreases to the value ξ_n satisfying $U'(\xi_n) = y^\infty \frac{q_n}{p_n}$. Similarly, we can prove that μ^N decreases to the value μ which satisfies $U'(\mu) = y^\infty r_n$, where $r_n = \frac{1-\sum_{k=1}^n q_k}{1-\sum_{k=1}^n p_k}$. In fact, by (3.37) it is sufficient to prove that r_n is increasing. To see this, we use the assumption $q_n < \frac{p_n}{2}$, which implies:

$$
r_{N+1} = \frac{1 - \sum_{n=1}^N q_n - q_{N+1}}{1 - \sum_{n=1}^N p_n - p_{N+1}} > \frac{1 - \sum_{n=1}^N q_n - \frac{p_{N+1}}{2}}{1 - \sum_{n=1}^N p_n - p_{N+1}} > \frac{1 - \sum_{n=1}^N q_n}{1 - \sum_{n=1}^N p_n} = r_N
$$

where the last inequality is equivalent to $1 - \sum_{n=1}^N q_n \geq \frac{1-\sum_{n=1}^N p_n}{2}$, which is implied by $q_n < \frac{p_n}{2}$.

We now show that $u(x) = \lim_{N\to\infty} u^N(x)$. Since $u^N(x)$ is increasing, and $u^N(x) \leq u(x)$ for all N, it is clear that $\lim_{N\to\infty} u^N(x) \leq u(x)$. To prove the reverse inequality, it suffices to show that $\widehat{X}_T^N$ is a maximizing sequence.

Consider a maximizing sequence for $u(x)$, that is a sequence $(Y^k)_{k=1}^\infty \subseteq \mathcal{X}(x)$ such that $\lim_{k\to\infty} \mathbf{E}[U(Y^k)] = u(x)$. We may well replace Y^k with $(1 - \delta_k)Y^k \mathbf{1}_{\{Y^k < M_k\}} + x\delta_k$ for some small $(\delta_k)_{k=1}^\infty$ and some big $(M_k)_{k=1}^\infty$, so that we still have a maximizing sequence (still denoted by $(Y^k)_{k=1}^\infty$), and Y_k is bounded away from zero.

As Y^k is the terminal payoff of some strategy H^k, we denote by Y_n^k the payoff of H^k at time n. Then we have that $U(Y_n^k)$ is bounded from below by $U(x\delta_k)$, uniformly in n. Therefore, Fatou's lemma implies that:

$$
\lim_{n\to\infty} \mathbf{E}[U(Y_n^k)] \geq \mathbf{E}[U(Y^k)]
$$

and hence, choosing n_k big enough, $(Y_{n_k}^k)_{k=1}^\infty$ is a maximizing sequence. On the other hand, we trivially have that $\mathbf{E}[U(Y_{n_k}^k)] \leq \mathbf{E}[U(\widehat{X}_T^k)]$, as desired.

We now prove that $u(x) < \infty$ for all $x > 0$. By Theorem 3.1 in [KS99] (or by the concavity of u), it suffices to show that $u(x) < \infty$ for some $x > 0$, and we show this for x^*. The utility maximization problem with initial capital x^* admits an optimizer (see also Lemma 3.10 (i) below), which is given by:

$$
\widehat{X}_T(x^*) = \sum_{n=1}^\infty \mathbf{1}_{A_n} I\left(y^\infty \frac{q_n}{p_n}\right) + \mathbf{1}_B I(y^\infty r_\infty).
$$

By assumption (e) above, $I\left(y^\infty \frac{q_n}{p_n}\right) \in \Xi$ for all n. Denoting by $M = U(I(y^\infty r_\infty))$, we have:

$$
\mathbf{E}[U(\widehat{X}_T(x^*))] < M + 2\mathbf{E}[\widehat{X}_T(x^*)U'(\widehat{X}_T(x^*))\mathbf{1}_{\Omega\setminus(B\cup A_1)}]
$$

$$
= M + 2y^\infty \sum_{n=2}^N I\left(y^\infty \frac{q_n}{p_n}\right) q_n = M + 2y^\infty x^*
$$

which shows that $u(x^*) < \infty$. $\square$

In the next lemma we describe the properties of the above model:

LEMMA 3.10. *The model constructed in Lemma 3.9 has the following properties*:
i) *For all $x \geq x^*$, $\widehat{X}_T^N(x)$ converges to $\widehat{X}_T(x^*)$ a.s.*
ii) *$u(x)$ is a straight line with slope $u'(x^*)$ for $x > x^*$.*
iii) *The utility maximization problem admits a solution if and only if $x \leq x^*$.*

PROOF.
i) Denoting by $\widehat{X}^N(x) = (\xi_1^N, \ldots, \xi_N^N, \mu^N)$, from (3.36), (3.37) we obtain that, for $x \geq x^*$:

$$\lim_{N \to \infty} \xi_n^N = I\left(y^N \frac{q_n}{p_n}\right) = I\left(y^\infty \frac{q_n}{p_n}\right)$$
$$\lim_{N \to \infty} \mu^N = I(y^N r_N) = I(y^\infty r_\infty)$$

therefore it suffices to show that this limit coincides with $\widehat{X}(x^*)$. In fact, from (3.35), we obtain that:

$$\sum_{n=1}^{N-1} I\left(y^N \frac{q_n}{p_n}\right) q_n + I(y^N r_N)\left(1 - \sum_{n=1}^{N} q_n\right) = x_N$$

and taking the limit as $N \to \infty$, we obtain:

$$\sum_{n=1}^{\infty} I\left(y^\infty \frac{q_n}{p_n}\right) q_n + I(y^\infty r_\infty)\left(1 - \sum_{n=1}^{\infty} q_n\right) = x^*$$

which concludes the proof.

Note that for $x < x^*$, equations (3.35), (3.36), (3.37) still admit a solution, but y^N converges to a value strictly greater than y^∞.

ii) By Lemma 3.9 we have that $u^N(x) \to u(x)$ for all $x > 0$, and by standard results on convex functions (see [R70]), this implies that also $(u^N)'(x)$ converges to $u'(x)$, for all $x > 0$ as $u(x)$ is differentiable by Theorem 2.0 in [KS99]. Also, by Theorem 2.0 in [KS99] we have that $(u^N)'(x) = y^N$. From i) it follows that, for $x \geq x^*$:

$$u'(x^*) = y^\infty = \lim_{N \to \infty} y^N = \lim_{N \to \infty} (u^N)'(x) = u'(x)$$

which proves the claim.

iii) As noted in i), an optimizer exists for $x = x^*$. Then an optimizer exists also for $x < x^*$ by Theorem 2.0 in [KS99]. Since an optimizer exists for $x = x^*$, then it cannot exist for $x > x^*$ by ii), and by the strict concavity of U (see Scholium 5.1 in [KS99]). $\square$

We now prove the technical lemma used in the proof of Proposition 3.8:

LEMMA 3.11. *Let U be a utility function satisfying Assumption 1.2 and (3.27). For all $\varepsilon > 0$, there is $x > \varepsilon^{-1}$ such that:*
  i) $xU'(x) \geq \varepsilon^{-1}$
  ii) $\frac{zU'(z)}{U(z)} \geq 1 - \varepsilon$ *for all $z \in [\varepsilon x, \frac{x}{\varepsilon}]$*
  iii) $U'(\varepsilon x) \leq (1 + \varepsilon)U'(x)$

Notice that ii) is an immediate consequence of (3.27), while iii) follows from Lemma 6.5 in [KS99]. The difficulty of the above lemma consists in finding some x which satisfies simultaneously ii) and iii).

The proof requires two auxiliary lemmata.

LEMMA 3.12. *Let U be a utility function satisfying Assumption 1.2, and $y < x$ such that:*

$$\frac{yU'(y)}{U(y)} \leq \alpha \qquad \frac{xU'(x)}{U(x)} \geq \beta.$$

Then we have that:

$$x \geq y\left(\frac{\beta - \alpha}{(1 - \beta)\alpha} + 1\right).$$

PROOF. Define $b(x) = U(x) - xU'(x)$. Note that $b(x)$ can be characterized as:

(3.42) $b(x) = \inf\{m : U'(x)z + m \geq U(z) \text{ for all } z > 0\}$

and therefore $b(x)$ is an increasing function of x (when U is twice differentiable, this is immediately seen by differentiating the definition of $b(x)$). This fact, combined with the assumptions on x and y, implies that:

(3.43) $(1 - \alpha)U(y) \leq b(y) \leq b(x) \leq (1 - \beta)U(x)$

and by the concavity of U:

(3.44) $U(x) \leq U(y) + U'(y)(x - y).$

Putting together (3.43) and (3.44), we have

$$x - y \geq \frac{\beta - \alpha}{1 - \beta}\frac{U(y)}{U'(y)}$$

and hence

$$x \geq y\left(\frac{\beta - \alpha}{1 - \beta}\frac{U(y)}{yU'(y)} + 1\right) \geq y\left(\frac{\beta - \alpha}{(1 - \beta)\alpha} + 1\right). \qquad \square$$

LEMMA 3.13. *Let U be a utility function satisfying Assumption 1.2, and $U(\infty) > 0$. Let $0 < y < x$ and $0 < \alpha < \beta$.*
  i) *If $\frac{xU'(x)}{U(x)} \geq \beta$ and $\frac{zU'(z)}{U(z)} \geq \alpha$ for all $z \in [y, x]$ then $U'(x) > \frac{\beta}{\alpha}\left(\frac{x}{y}\right)^{\alpha-1}U'(y)$*
  ii) *For y large enough, if $\frac{yU'(y)}{U(y)} \geq \beta$ and $\frac{xU'(x)}{U(x)} \leq \alpha$ then $U'(y) \geq \frac{\beta}{\alpha}U'(x)$*

PROOF.

i) As in the proof of Lemma 3.4, we have

$$U(z) \geq U(y) \left(\frac{z}{y}\right)^{\alpha} \quad \text{for all} \quad z \in [y, x].$$

It follows that

$$\beta \leq \frac{xU'(x)}{U(x)} \leq \frac{yU'(y)}{U(y)} \left(\frac{y}{x}\right)^{\alpha-1} \leq \alpha \frac{U'(x)}{U'(y)} \left(\frac{y}{x}\right)^{\alpha-1}$$

which completes the proof.

ii) Notice that

$$\left(\frac{U(x)}{x}\right)' = \frac{xU'(x) - U(x)}{x^2} = -\frac{b(x)}{x^2}.$$

Since $U(\infty) > 0$ implies that $\lim_{x \to \infty} b(x) > 0$, it follows that $\frac{U(x)}{x}$ is a decreasing positive function on the nonempty set $\{x : b(x) \geq 0, U(x) \geq 0\}$. It follows that:

$$\frac{U'(y)}{U'(x)} \geq \frac{\beta}{\alpha} \frac{U(y)}{y} \frac{x}{U(x)} \geq \frac{\beta}{\alpha}$$

as claimed. $\qquad\square$

PROOF OF LEMMA 3.11. First note that $AE_{+\infty}(U) > 0$ implies that U is unbounded (Lemma 6.1 in [KS99]). Hence, we only need to find some x big enough which satisfies ii) and iii), as i) will follow automatically.

If there exists some x_0 such that $\frac{xU'(x)}{U(x)} \geq 1 - \varepsilon$ for all $x \geq x_0$, the result follows by Lemma 6.5 in [KS99]. In fact, from the mentioned lemma we obtain that, for all $\alpha < 1$ and $\beta > 1$ there exists $y > \varepsilon^{-1}$ such that:

$$\alpha U'(y) < U'(\beta y)$$

and iii) follows by setting $\beta = \frac{1}{\varepsilon}$, $x = \frac{y}{\varepsilon}$ and $\alpha = \frac{1}{1+\varepsilon}$. Such an x will then satisfy ii) and i) above.

Otherwise, if $\liminf_{x \to \infty} \frac{xU'(x)}{U(x)} < 1$, then there exist $y, \hat{x}$, with $y < \hat{x}$ such that:

$$\frac{\hat{x}U'(\hat{x})}{U(\hat{x})} \geq 1 - \varepsilon^3, \quad \frac{yU'(y)}{U(y)} = 1 - \varepsilon \quad \text{and} \quad \frac{zU'(z)}{U(z)} \geq 1 - \varepsilon \quad \text{for all} \quad z \in [y, \hat{x}].$$

In fact, we may just choose some $y < \hat{x}$ with the first two properties, and then replace y with $\sup\{w < \hat{x} : \frac{wU'(w)}{U(w)} \leq 1 - \varepsilon\}$.

Denote now $x = \inf\{w > y : \frac{wU'(w)}{U(w)} = 1 - \varepsilon^2\}$. Notice that, by Lemma 3.12, we have:

$$\hat{x} \geq \left(1 + \frac{1}{\varepsilon(1+\varepsilon)}\right) x \geq \frac{x}{\varepsilon} \quad \text{and} \quad \varepsilon x \geq y(1+\varepsilon) > y.$$

Applying Lemma 3.13 (i) to $y = \varepsilon x$, we obtain that:

$$U'(x) \geq (1+\varepsilon)\varepsilon^{\varepsilon} U'(\varepsilon x) > (1+\varepsilon)(1+\varepsilon \log \varepsilon)U'(\varepsilon x)$$

and iii) follows for ε small enough. $\qquad\square$

3.3. – Existence theorems

Let us now move to the positive results in the spirit of Theorem 2.16 and Theorem 2.18 above. We first consider the case where U satisfies Case 1 of Assumption 1.2, which was studied in [KS99].

CASE 1. $\mathrm{dom}(U) = \mathbb{R}_+$.

The heart of the argument in the proof of Theorem 2.18 (which we now want to extend to the general case) is to find a saddlepoint for the Lagrangian. In more general situations we have to apply the minimax theorem, which is crucial in the theory of Lagrange multipliers. We want to extend the applicability of the minimax theorem to the present situation. The infinite-dimensional versions of the minimax theorem available in the literature (see, e.g, [ET76] or [St85]) are along the following lines: let $\langle E, F \rangle$ be a pair of locally convex vector spaces in separating duality, $C \subseteq E$, $D \subseteq F$ a pair of convex subsets, and $L(x, y)$ a function defined on $C \times D$, concave in the first and convex in the second variable, having some (semi-)continuity property compatible with the topologies of E and F (which in turn should be compatible with the duality between E and F). If (at least) one of the sets C and D is compact and the other is complete, then one may assert the existence of a saddle point $(\widehat{\xi}, \widehat{\eta}) \in C \times D$ such that

$$(3.45) \qquad L(\widehat{\xi}, \widehat{\eta}) = \sup_{\xi \in C} \inf_{\eta \in D} L(\xi, \eta) = \inf_{\eta \in D} \sup_{\xi \in C} L(\xi, \eta).$$

We try to apply this theorem to the analogue of the Lagrangian encountered in the proof of Theorem 2.18 above. Fixing $x > 0$ and $y > 0$ let us formally write the Lagrangian (2.70) in the infinite-dimensional setting,

$$(3.46) \qquad L^{x,y}(X_T, \mathbf{Q}) = \mathbf{E_P}[U(X_T)] - y(\mathbf{E_Q}[X_T - x])$$

$$(3.47) \qquad = \mathbf{E_P}\left[U(X_T) - y\frac{d\mathbf{Q}}{d\mathbf{P}}X_T \right] + yx,$$

where X_T runs through "all" non-negative $\mathcal{F}_T$-measurable functions and $\mathbf{Q}$ through the set $\mathcal{M}^a(S)$ of absolutely continuous local martingale measures.

To restrict the set of "all" nonnegative functions to a more amenable one, note that $\inf_{y>0, \mathbf{Q} \in \mathcal{M}^a(S)} L^{x,y}(X_T, \mathbf{Q}) > -\infty$ iff

$$(3.48) \qquad \mathbf{E_Q}[X_T] \leq x, \quad \text{for all} \quad \mathbf{Q} \in \mathcal{M}^a(S).$$

Using the basic result on the super-replicability of the contingent claim X_T (see [KQ95], [J92], [AS94], [DS94], and [DS98]), we have – as encountered in Theorem 2.11 for the finite dimensional case – that a non-negative $\mathcal{F}_T$-measurable random variable X_T satisfies (3.48) iff there is an admissible trading strategy H such that

$$(3.49) \qquad X_T \leq x + (H \cdot S)_T.$$

Hence let

$$C(x) = \{X_T \in L_+^0(\Omega, \mathcal{F}_T, \mathbf{P}) :$$

(3.50)
$$X_T \leq x + (H \cdot S)_T, \text{ for some admissible } H\}$$
$$= \{X_T \in L_+^0(\Omega, \mathcal{F}_T, \mathbf{P}) :$$

(3.51)
$$\mathbf{E}_\mathbf{Q}[X_T] \leq x, \text{ for all } \mathbf{Q} \in \mathcal{M}^a(S)\}$$

and simply write C for $C(1)$ (observe that $C(x) = xC$).

We thus have found a natural set $C(x)$ in which X_T should vary when we are mini-maxing the Lagrangian $L^{x,y}$. Dually, the set $\mathcal{M}^a(S)$ seems to be the natural domain where the measure $\mathbf{Q}$ is allowed to vary (in fact, we shall see later, that this set still has to be slightly enlarged). But what are the locally convex vector spaces E and F in separating duality into which C and $\mathcal{M}^a(S)$ are naturally embedded? As regards $\mathcal{M}^a(S)$ the natural choice seems to be $L^1(\mathbf{P})$ (by identifying a measure $\mathbf{Q} \in \mathcal{M}^a(S)$ with its Radon-Nikodym derivative $\frac{d\mathbf{Q}}{d\mathbf{P}}$); note that $\mathcal{M}^a(S)$ is a *closed* subset of $L^1(\mathbf{P})$, which is good news. On the other hand, there is no reason for C to be contained in $L^\infty(\mathbf{P})$, or even in $L^p(\mathbf{P})$, for any $p > 0$; the natural space in which C is embedded is just $L^0(\Omega, \mathcal{F}_T, \mathbf{P})$, the space of all real-valued $\mathcal{F}_T$-measurable functions endowed with the topology of convergence in probability.

The situation now seems hopeless (if we don't want to impose artificial $\mathbf{P}$-integrability assumptions on X_T and/or $\frac{d\mathbf{Q}}{d\mathbf{P}}$), as $L^0(\mathbf{P})$ and $L^1(\mathbf{P})$ are not in any reasonable duality; in fact, $L^0(\mathbf{P})$ is not even a locally convex space, hence there seems to be no hope for a good duality theory, which could serve as a basis for the application of the minimax theorem. But the good news is that the sets C and $\mathcal{M}^a(S)$ are in the *positive orthant* of $L^0(\mathbf{P})$ and $L^1(\mathbf{P})$ respectively; the crucial observation is, that for $f \in L_+^0(\mathbf{P})$ and $g \in L_+^1(\mathbf{P})$, it is possible to well-define

(3.52)
$$\langle f, g \rangle := \mathbf{E}_\mathbf{P}[fg] \in [0, \infty].$$

The spirit here is similar as in the very foundation of Lebesgue integration theory: for positive measurable functions the integral is always well-defined, but possibly $+\infty$. This does not cause any logical inconsistency.

Similarly the bracket $\langle .,. \rangle$ defined in (3.52) shares many of the usual properties of a scalar product. The difference is that $\langle f, g \rangle$ now may assume the value $+\infty$ and that the map $(f, g) \mapsto \langle f, g \rangle$ is not continuous on $L_+^0(\mathbf{P}) \times L_+^1(\mathbf{P})$, but only lower semi-continuous (this immediately follows from Fatou's lemma).

At this stage it becomes clear that the role of $L_+^1(\mathbf{P})$ is somewhat artificial, and it is more natural to define (3.52) in the general setting where f and g are both allowed to vary in $L_+^0(\mathbf{P})$. The pleasant feature of the space $L^0(\mathbf{P})$ in the context of Mathematical Finance is, that it is invariant under the passage to an equivalent measure $\mathbf{Q}$, a property only shared by $L^\infty(\mathbf{P})$, but by no other $L^p(\mathbf{P})$, for $0 < p < \infty$.

We now can turn to the polar relation between the sets C and $\mathcal{M}^a(S)$. By (3.49) we have, for an element $X_T \in L^0_+(\Omega, \mathcal{F}, \mathbf{P})$,

$$(3.53) \qquad X_T \in C \Longleftrightarrow \mathbf{E}_\mathbf{Q}[X_T] = \mathbf{E}_\mathbf{P}\left[X_T \frac{d\mathbf{Q}}{d\mathbf{P}}\right] \le 1, \quad \text{for} \quad \mathbf{Q} \in \mathcal{M}^a(S).$$

Denote by D the closed, convex, solid hull of $\mathcal{M}^a(S)$ in $L^0_+(\mathbf{P})$. It is easy to show (using, e.g., Lemma 3.14 below), that D equals

$$(3.54) \qquad D = \left\{ Y_T \in L^0_+(\Omega, \mathcal{F}_T, \mathbf{P}) : \text{ there is} \right.$$
$$\left. (\mathbf{Q}_n)^\infty_{n=1} \in \mathcal{M}^a(S) \text{ s.t. } Y_T \le \lim_{n \to \infty} \frac{d\mathbf{Q}_n}{d\mathbf{P}} \right\},$$

where the $\lim_{n\to\infty} \frac{d\mathbf{Q}_n}{d\mathbf{P}}$ is understood in the sense of almost sure convergence. We have used the letter Y_T for the elements of D to stress the dual relation to the elements X_T in C. In further analogy we write, for $y > 0$, $D(y)$ for yD, so that $D = D(1)$. By (3.54) and Fatou's lemma we again find that, for $X_T \in L^0_+(\Omega, \mathcal{F}, \mathbf{P})$

$$(3.55) \qquad X_T \in C \Longleftrightarrow \mathbf{E}_\mathbf{P}[X_T Y_T] \le 1, \quad \text{for} \quad Y_T \in D.$$

Why did we pass to this enlargement D of the set $\mathcal{M}^a(S)$? The reason is that we now obtain a more symmetric relation between C and D: for $Y_T \in L^0_+(\Omega, \mathcal{F}, \mathbf{P})$ we have

$$(3.56) \qquad Y_T \in D \Longleftrightarrow \mathbf{E}_\mathbf{P}[X_T Y_T] \le 1, \quad \text{for} \quad X_T \in C.$$

The proof of (3.56) relies on an adaption of the "bipolar theorem" from the theory of locally convex spaces (see, e.g., [S66]) to the present duality $\langle L^0_+(\mathbf{P}), L^0_+(\mathbf{P}) \rangle$, which was worked out in [BS99].
    Why is it important to define the enlargement D of $\mathcal{M}^a(S)$ in such a way that (3.56) holds true? After all, $\mathcal{M}^a(S)$ is a nice, convex, closed (w.r.t. the norm of $L^1(\mathbf{P})$) set and one may prove that, for $g \in L^1(\mathbf{P})$ *such that* $\mathbf{E}_\mathbf{P}[g] = 1$,

$$(3.57) \qquad g \in \mathcal{M}^a(S) \Longleftrightarrow \mathbf{E}_\mathbf{P}[X_T g] \le 1, \quad \text{for} \quad X_T \in C.$$

The reason is that, in general, the saddle point $(\widehat{X}_T, \widehat{\mathbf{Q}})$ of the Lagrangian will *not* be such that $\widehat{\mathbf{Q}}$ is a probability measure; it will only satisfy $\mathbf{E}[\frac{d\widehat{\mathbf{Q}}}{d\mathbf{P}}] \le 1$, the inequality possibly being strict. But it will turn out that $\widehat{\mathbf{Q}}$, which we identify with $\frac{d\widehat{\mathbf{Q}}}{d\mathbf{P}}$, is always in D. In fact, the passage from $\mathcal{M}^a(S)$ to D is the *crucial feature* in order to make the duality work in the present setting: we shall see below that even for nice utility functions U, such as the logarithm, and for nice processes, such as a continuous process $(S_t)_{0 \le t \le T}$ based on the filtration of two Brownian motions, the above described phenomenon can occur: the saddle point of the Lagrangian leads out of $\mathcal{M}^a(S)$.

The set D can be characterized in several equivalent manners. We have defined D above in the abstract way as the convex, closed, solid hull of $\mathcal{M}^a(S)$ and mentioned the description (3.54). Equivalently, one may define D as the set of random variables $Y_T \in L^0_+(\Omega, \mathcal{F}, \mathbf{P})$ such that there is a process $(Y_t)_{0 \le t \le T}$ starting at $Y_0 = 1$ with $(Y_t X_t)_{0 \le t \le T}$ a $\mathbf{P}$-supermartingale, for every non-negative process $(X_t)_{0 \le t \le T} = (x + (H \cdot S)_t)_{0 \le t \le T}$, where $x > 0$ and H is predictable and S-integrable. This definition was used in [KS99]. Another equivalent characterization was used in [CSW01]: consider the convex, solid hull of $\mathcal{M}^a(S)$, and embed this subset of $L^1(\mathbf{P})$ into the bidual $L^1(\mathbf{P})^{**} = L^\infty(\mathbf{P})^*$; denote by $\overline{\mathcal{M}^a(S)}$ the weak-star closure of the convex solid hull of $\mathcal{M}^a(S)$ in $L^\infty(\mathbf{P})^*$. Each element of $\overline{\mathcal{M}^a(S)}$ may be decomposed into its regular part $\mu^r \in L^1(\mathbf{P})$ and its purely singular part $\mu^s \in L^\infty(\mathbf{P})^*$. It turns out that D equals the set $\{\mu^r \in L^1(\mathbf{P}) : \mu \in \overline{\mathcal{M}^a(S)}\}$, i.e. consists of the regular parts of the elements of $\overline{\mathcal{M}^a(S)}$. This description has the advantage that we may associate to the elements $\mu^r \in D$ a singular part μ^s, and it is this extra information which is crucial when extending the present results to the case of random endowment as in [CSW 01]. Compare also [HK02], where the case of random endowment is analyzed in full generality without using the space $L^\infty(\mathbf{P})^*$.

Why are the sets C and D hopeful candidates for the minimax theorem to work out properly for a function L defined on $C \times D$? Both are closed, convex and bounded subsets of $L^0_+(\mathbf{P})$. But recall that we still need some compactness property to be able to localize the mini-maximizers (resp. maxi-minimizers) on C (resp. D). In general, neither C nor D is compact (w.r.t. the topology of convergence in measure), i.e., for a sequence $(f_n)_{n=1}^\infty$ in C (resp. $(g_n)_{n=1}^\infty$ in D) we cannot pass to a subsequence converging in measure. But C and D have a property which is close to compactness and in many applications turns out to serve just as well.

LEMMA 3.14. *Let A be a closed, convex, bounded subset of $L^0_+(\Omega, \mathcal{F}, \mathbf{P})$. Then for each sequence $(h_n)_{n=1}^\infty \in A$ there exists a sequence of convex combinations $k_n \in \mathrm{conv}(h_n, h_{n+1}, \dots)$ which converges almost surely to a function $k \in A$.*

This easy lemma (see, e.g., [DS94, Lemma A.1.1], for a proof) is in the spirit of the celebrated theorem of Komlos [Kom67], stating that for a bounded sequence $(h_n)_{n=1}^\infty$ in $L^1(\mathbf{P})$ there is a subsequence converging in Cesaro-mean almost surely. The methodology of finding pointwise limits by using convex combinations has turned out to be extremely useful as a surrogate for compactness. For an extensive discussion of more refined versions of the above lemma and their applications to Mathematical Finance we refer to [DS99].

The application of the above lemma is the following: by passing to convex combinations of optimizing sequences $(f_n)_{n=1}^\infty$ in C (resp $(g_n)_{n=1}^\infty$ in D), we can always find limits $f \in C$ (resp. $g \in D$) w.r.t. almost sure convergence. Note that the passage to convex combinations does not cost more than passing to a subsequence in the application to convex optimization.

We have now given sufficient motivation to state the central result of [KS99], which is the generalization of Theorem 2.18 to the semi-martingale setting under Assumption 1.2, Case 1, and having reasonable asymptotic elasticity.

THEOREM 3.15 ([KS99], Theorem 2.2). *Let the semi-martingale* $S = (S_t)_{0 \le t \le T}$ *and the utility function U satisfy Assumptions 1.1, 1.2 Case 1 and 1.3; suppose in addition that U has reasonable asymptotic elasticity. Define*

$$(3.58) \qquad u(x) = \sup_{X_T \in C(x)} E[U(X_T)], \quad v(y) = \inf_{Y_T \in D(y)} E[V(Y_T)].$$

Then we have:

(i) *The value functions* $u(x)$ *and* $v(y)$ *are conjugate; they are continuously differentiable, strictly concave (resp. convex) on* $]0, \infty[$ *and satisfy*

$$(3.59) \qquad u'(0) = -v'(0) = \infty, \quad u'(\infty) = v'(\infty) = 0.$$

(ii) *The optimizers* $\widehat{X}_T(x)$ *and* $\widehat{Y}_T(y)$ *in (3.58) exist, are unique and satisfy*

$$(3.60) \qquad \widehat{X}_T(x) = I(\widehat{Y}_T(y)), \quad \widehat{Y}_T(y) = U'(\widehat{X}_T(x)),$$

where $x > 0$, $y > 0$ *are related via* $u'(x) = y$ *or equivalently* $x = -v'(y)$.

(iii) *We have the following relations between* u', v' *and* $\widehat{X}_T$, $\widehat{Y}_T$ *respectively:*

$$(3.61) \qquad \begin{aligned} u'(x) &= E\left[\frac{\widehat{X}_T(x)U'(\widehat{X}_T(x))}{x}\right], \quad x > 0, \\ v'(y) &= E\left[\frac{\widehat{Y}_T(y)V'(\widehat{Y}_T(y))}{y}\right], \quad y > 0. \end{aligned}$$

For the full proof of the theorem we refer to [KS99]. Instead, here we present an alternative proof of the existence of the solution $\widehat{X}_T(x)$, which does not rely on the dual point of view.

We start with an easy lemma, which is in the spirit of Kadec' and Pełczyn'ski [KP 65]:

LEMMA 3.16. *Let* $(f_n)_{n=1}^{\infty} \ge 0$ *be random variables on* $(\Omega, \mathcal{F}, P)$ *converging a.s. to* f_0. *Suppose that* $\lim_{n \to \infty} E[f_n] = E[f_0] + \alpha$, *for some* $\alpha > 0$. *Then for all* $\varepsilon > 0$ *there exist* $n, m > \varepsilon^{-1}$ *and disjoint sets* A_n, A_m *such that the following conditions are satisfied:*

i) $f_n \ge \varepsilon^{-1}$ *on* A_n *and* $f_m \ge \varepsilon^{-1}$ *on* A_m

ii) $E[f_n 1_{A_n}] > \alpha - \varepsilon$ *and* $E[f_m 1_{A_m}] > \alpha - \varepsilon$

iii) $E[f_n 1_{\Omega \setminus (A_n \cup A_m)}] > E[f_0] - \varepsilon$ *and* $E[f_m 1_{\Omega \setminus (A_n \cup A_m)}] > E[f_0] - \varepsilon$.

PROOF. Denoting by $B_k = \{f_k \geq (f_0 + \varepsilon) \vee \varepsilon^{-1}\}$, we can write:

$$\mathbf{E}[f_k] = \mathbf{E}[f_k \mathbf{1}_{\Omega \setminus B_k}] + \mathbf{E}[f_k \mathbf{1}_{B_k}].$$

Since the class $(f_k \mathbf{1}_{\Omega \setminus B_k})_{k \in \mathbb{N}}$ is uniformly integrable, the first term in the right converges to $\mathbf{E}[f_0]$, and therefore the last term converges to α. This means that we can choose $n, m > \varepsilon^{-1}$ such that:

$$\mathbf{E}[f_n \mathbf{1}_{B_n}] > \alpha - \frac{\varepsilon}{2} \quad \text{and} \quad \mathbf{E}[f_n \mathbf{1}_{\Omega \setminus B_n}] > \mathbf{E}[f_0] - \frac{\varepsilon}{2}$$

$$\mathbf{E}[f_m \mathbf{1}_{B_m}] > \alpha - \frac{\varepsilon}{2} \quad \text{and} \quad \mathbf{E}[f_m \mathbf{1}_{\Omega \setminus B_m}] > \mathbf{E}[f_0] - \frac{\varepsilon}{2}.$$

The uniform integrability of $(f_k \mathbf{1}_{\Omega \setminus B_k})_{k \in \mathbb{N}}$ also implies that for any $\varepsilon > 0$ we can find $\delta > 0$ such that, if $\mathbf{P}(C) < \delta$, then:

(3.62)
$$\mathbf{E}[f_k \mathbf{1}_{C \setminus B_k}] = \mathbf{E}[f_k \mathbf{1}_{\Omega \setminus B_k} \mathbf{1}_C] < \varepsilon.$$

Since f_k converges a.s. to f_0, for all $\varepsilon > 0$ we have:

$$\lim_{k \to \infty} \mathbf{P}(B_k) = 0$$

and hence we can choose B_n small enough, so that by (3.62), $\mathbf{E}[f_k \mathbf{1}_{B_n \setminus B_k}] < \frac{\varepsilon}{2}$ for all k. Analogously, we choose m such that $\mathbf{E}[f_n \mathbf{1}_{B_m}] < \frac{\varepsilon}{2}$.

Denoting by $A_n = B_n \setminus B_m$ and $A_m = B_m$, we obviously have $A_n \cap A_m = \emptyset$, and it is easy to check that they satisfy i), ii), and iii). □

PROOF OF EXISTENCE OF THE PRIMAL OPTIMIZER $X_T(x)$. Fix $x > 0$ and let $X_n(x) = x + (H_n \cdot S)_T \geq 0$ be a maximizing sequence for (1.16). Since $\mathrm{AE}_{+\infty}(U) < 1$, by Lemma 6.3 in [KS99] there exists some $\gamma > 1$ such that $U(\frac{x}{2}) > \frac{\gamma}{2} U(x)$ for all $x \geq x_0$.

By passing to a sequence of convex combinations $X'_n \in \mathrm{conv}(X_n, X_{n+1}, \dots)$, still denoted by X_n and applying Lemma 3.14, we may suppose that X_n converges a.s. to $\widehat{X} \in L^0(\Omega, \mathcal{F}, \mathbf{P}, [0, \infty])$. Since $\mathcal{M}^e(S) \neq \emptyset$, we have that $\mathrm{conv}(X_n, X_{n+1}, \dots)$ is bounded in $L^1(\mathbf{Q})$, and hence in $L^0(\mathbf{P})$ (that is, bounded in probability under $\mathbf{P}$). Hence $\widehat{X} < \infty$ is finite almost surely, and by the closedness of $C(x)$ in $L^0(\mathbf{P})$ we have $X_0 \leq x + (\widehat{H} \cdot S)_T$ for some admissible $\widehat{H}$.

We shall show that $\lim_{n \to \infty} \mathbf{E}[U(X_n)] = \mathbf{E}[U(\widehat{X})]$. If not, we have that:

$$\lim_{n \to \infty} \mathbf{E}[U(X_n)] - \mathbf{E}[U(\widehat{X})] = \alpha > 0$$

and by Lemma 3.16 we may find n, m, A_n, A_m such that:

$$U(X_n) \geq \varepsilon^{-1} \quad \text{on} \quad A_n \qquad\qquad U(X_m) \geq \varepsilon^{-1} \quad \text{on} \quad A_m$$

$$\mathbf{E}[U(X_n)\mathbf{1}_{A_n}] > \alpha - \varepsilon \qquad\qquad \mathbf{E}[U(X_m)\mathbf{1}_{A_m}] > \alpha - \varepsilon$$

$$\mathbf{E}[U(X_n)\mathbf{1}_{\Omega \setminus (A_n \cup A_m)}] > \mathbf{E}[U(\widehat{X})] - \varepsilon \quad \mathbf{E}[U(X_m)\mathbf{1}_{\Omega \setminus (A_n \cup A_m)}] > \mathbf{E}[U(\widehat{X})] - \varepsilon.$$

We can write:

$$\mathbf{E}\left[U\left(\frac{X_n + X_m}{2}\right)\right] = \mathbf{E}\left[U\left(\frac{X_n + X_m}{2}\right)\mathbf{1}_{\Omega\setminus(A_n\cup A_m)}\right]$$
$$+ \mathbf{E}\left[U\left(\frac{X_n + X_m}{2}\right)\mathbf{1}_{A_n\cup A_m}\right].$$

By the condition $AE_{+\infty(U)} < 1$, for the second term we have:

$$\mathbf{E}\left[U\left(\frac{X_n + X_m}{2}\right)\mathbf{1}_{A_n\cup A_m}\right] \geq \gamma\mathbf{E}\left[\frac{U(X_n + X_m)}{2}\mathbf{1}_{A_n\cup A_m}\right]$$
$$\geq \frac{\gamma}{2}\left(\mathbf{E}[U(X_n)\mathbf{1}_{A_n}] + \mathbf{E}[U(X_m)\mathbf{1}_{A_m}]\right)$$
$$\geq \gamma(\alpha - \varepsilon)$$

while for the first term:

$$\mathbf{E}\left[U\left(\frac{X_n + X_m}{2}\right)\mathbf{1}_{A_n\cup A_m}\right]$$
$$\geq \frac{1}{2}\left(\mathbf{E}[U(X_n)\mathbf{1}_{\Omega\setminus(A_n\cup A_m)}] + \mathbf{E}[U(X_m)\mathbf{1}_{\Omega\setminus(A_n\cup A_m)}]\right) \geq \mathbf{E}[U(\widehat{X})] - \varepsilon$$

and hence:

$$\mathbf{E}\left[U\left(\frac{X_n + X_m}{2}\right)\mathbf{1}_{A_n\cup A_m}\right] \geq \mathbf{E}[U(\widehat{X})] + \alpha + ((\gamma - 1)\alpha - \varepsilon(\gamma + 1)).$$

Since ε can be chosen arbitrarily small, we can assume that the last term in the right is positive, but this leads to a contradiction, since X_n was a maximizing sequence, and $\lim_{n\to\infty}\mathbf{E}[U(X_n)] = \mathbf{E}[U(\widehat{X})] + \alpha$ was the supremum. □

We finish the discussion of utility functions satisfying the Inada conditions (1.7) and (1.8) by briefly indicating two examples, when the dual optimizer $\widehat{Y}_T(y)$ fails to be of the form $\widehat{Y}_T(y) = y\frac{d\widehat{Q}(y)}{d\mathbf{P}}$, for some probability measure $\widehat{Q}(y)$.

The first example will feature a stopped Geometric Brownian Motion as the asset process, while the second one will be a simple one-period model with countably many states.

EXAMPLE 3.17 ([KS99], Example 5.2). It suffices to consider a stock-price process of the form

(3.63)
$$S_t = \left(\exp\left(B_t + \frac{t}{2}\right)\right)^{\tau}$$
$$= \exp\left(B_{t\wedge\tau} + \frac{t\wedge\tau}{2}\right), \quad t \geq 0,$$

where $(B_t)_{t\geq 0}$ is Brownian motion based on $(\Omega, \mathcal{F}, (\mathcal{F}_t)_{t>0}, \mathbf{P})$ and τ a suitably chosen finite stopping time (to be discussed below) with respect to the filtration $(\mathcal{F}_t)_{t>0}$, after which the process S remains constant.

The usual way to find a risk-neutral measure $\mathbf{Q}$ for the process S above is to use Girsanov's formula, which amounts to considering

$$(3.64) \qquad Z_\tau = \exp\left(-B_\tau - \frac{\tau}{2}\right)$$

as a candidate for the Radon-Nikodym derivative $\frac{d\mathbf{Q}}{d\mathbf{P}}$.

We shall construct τ in such a way that – apart from other properties to be discussed below – the density process given by Girsanov's theorem

$$(3.65) \qquad Z_t = \exp\left(-B_{t\wedge\tau} - \frac{t\wedge\tau}{2}\right), t > 0$$

fails to be a uniformly integrable martingale, so that

$$(3.66) \qquad \mathbf{E}[Z_\tau] < 1.$$

The trick is to choose the filtration $(\mathcal{F}_t)_{t\geq 0}$ to be generated by two independent Brownian motions $(B_t)_{t\geq 0}$ and $(W_t)_{t\geq 0}$. Using the information of *both* $(B_t)_{t\geq 0}$ and $(W_t)_{t\geq 0}$ one may define τ in a suitable way such that (3.66) holds true and nevertheless we have that $\mathcal{M}^e(S) \neq \emptyset$. In other words, there are equivalent martingale measures $\mathbf{Q}$ for the process S, but Girsanov's theorem fails to produce one.

This example is known for quite some time ([DS98]) and served as a kind of "universal counterexample" to several questions arising in Mathematical Finance.

How can one use this example in the present context? Consider the logarithmic utility $U(x) = \ln(x)$ and recall that its conjugate function V equals $V(y) = -\ln(y) - 1$. Hence the dual optimization problem – formally – is given by

$$(3.67) \qquad \begin{aligned} \mathbf{E}\left[V\left(y\frac{d\mathbf{Q}}{d\mathbf{P}}\right)\right] &= \mathbf{E}\left[-\ln\left(y\frac{d\mathbf{Q}}{d\mathbf{P}}\right) - 1\right] \\ &= -\mathbf{E}\left[\ln\left(\frac{d\mathbf{Q}}{d\mathbf{P}}\right)\right] - (\ln(y)+1) \longrightarrow \min!, \\ \mathbf{Q} &\in \mathcal{M}^a(S). \end{aligned}$$

It is well known (see, e.g., the literature on the "numéraire portfolio" [L90], [A97] and [B01]), that for a process $(S_t)_{t\geq 0}$ based, e.g., on the filtration generated by an n-dimensional Brownian motion, the martingale measure obtained from applying Girsanov's theorem (which equals the "minimal martingale measure" investigated by Föllmer and Schweizer [FS91]) is the minimizer for (3.67), *provided it exists.*

In the present example we have seen that the candidate for the density of the minimal martingale measure Z_τ obtained from a formal application of Girsanov's theorem fails to have full measure; but nevertheless one may show that Z_τ is the optimizer of the dual problem (3.63), which shows in particular that we have to pass from $\mathcal{M}^a(S)$ to the larger set D to find the dual optimizer in (3.67). $\square$

EXAMPLE 3.18. ([KS99], Example 5.2 bis]). Let $(p_n)_{n=0}^\infty$ be a sequence of strictly positive numbers, $\sum_{n=0}^\infty p_n = 1$, tending sufficiently fast to zero and $(x_n)_{n=0}^\infty$ a sequence of positive reals, $x_0 = 2$, decreasing also to zero (but less fast than $(p_n)_{n=0}^\infty$). For example, $p_0 = 1 - \alpha$, $p_n = \alpha 2^{-n}$, for $n \geq 1$, and $x_0 = 2$, $x_n = \frac{1}{n}$, for $n \geq 1$, will do, if $0 < \alpha < 1$ is small enough to satisfy $(1 - \alpha)/2 + \alpha \sum_{n=1}^\infty 2^{-n}(-n + 1) > 0$.

Now define $S \triangleq (S_0, S_1)$ by letting $S_0 \equiv 1$ and S_1 to take the values $(x_n)_{n=0}^\infty$ with probability p_n. As filtration we choose the natural filtration generated by S. Clearly the process S satisfies $\mathcal{M}^e(S) \neq \emptyset$. In this case we can explicitly calculate the family of admissible processes starting at 1: it consists of all processes X with $X_0 = 1$ and such that X_1 is equals the random variable $X^\lambda = 1 + \lambda(S_1 - S_0)$, for some $-1 \leq \lambda \leq 1$.

Using again $U(x) = \ln(x)$ as utility function and writing $f(\lambda) = E[U(X^\lambda)]$ we obtain by an elementary calculation

$$f'(\lambda) = \sum_{n=0}^\infty p_n \frac{x_n - 1}{1 + \lambda(x_n - 1)}$$

so that $f'(\lambda)$ is strictly positive for $-1 \leq \lambda \leq 1$ if $\alpha > 0$ satisfies the above assumption $f'(1) = (1 - \alpha)\frac{1}{2} + \alpha \sum_{n=1}^\infty 2^{-n}(-n + 1) > 0$. Hence $f(\lambda)$ attains its maximum on $[-1, 1]$ at $\lambda = 1$, in other words, the optimal investment process $\widehat{X}(1)$ equals the process S.

We can also explicitly calculate $u(x)$ by

$$(3.68) \qquad u(x) = \mathbf{E}[U(x S_1)] = \sum_{n=0}^\infty p_n U(x x_n)$$

$$(3.69) \qquad = \sum_{n=0}^\infty p_n (\ln(x) + \ln(x_n)) = \ln(x) + \sum_{n=0}^\infty p_n \ln(x_n).$$

In particular, $u'(1) = 1$ and by Theorem 3.15 (ii) we get $\widehat{Y}(1) = U'(\widehat{X}(1)) = (S_1)^{-1}$. Note that

$$\mathbf{E}[S_1^{-1}] = \sum_{n=0}^\infty \frac{p_n}{x_n} = \frac{p_0}{2} + \sum_{n=1}^\infty n p_n$$

is strictly less than 1 by using again the condition $(1 - \alpha)\frac{1}{2} + \alpha \sum_{n=1}^\infty 2^{-n} (-n + 1) > 0$. In particular, the optimal element $\widehat{Y}_1(1)$ is not the density of a martingale measure for the process S.

Passing again to the general setting of Theorem 3.15 one might ask: how severe is the fact encountered in the two examples above that the dual optimizer $\widehat{Y}_T(1)$ may fail to be the density of a probability measure (or that $\mathbf{E}[\widehat{Y}_T(y)] < y$, for $y > 0$, which amounts to the same thing)? In fact, in many respects it does not bother us at all: we still have the basic duality relation between the primal and the dual optimizer displayed in Theorem 3.15 (ii). Even more is true: using the terminology from [KS99] the product $(\widehat{X}_t(x)\widehat{Y}_t(y))_{0 \le t \le T}$, where x and y satisfy $u'(x) = y$, is a uniformly integrable martingale. This fact can be interpreted in the following way: by taking the optimal portfolio $(\widehat{X}_t(x))_{0 \le t \le T}$ as numéraire instead of the original cash account, the pricing rule obtained from the dual optimizer $\widehat{Y}_T(y)$ then is induced by an equivalent martingale measure. We refer to ([KS99], p. 912) for a thorough discussion of this argument.

Finally we want to draw the attention of the reader to the fact that – comparing item (iii) of Theorem 3.15 to the corresponding item of Theorem 2.18 – we only asserted one pair of formulas for $u'(x)$ and $v'(y)$. The reason is that, in general, the formulae (2.82) do not hold true any more, the reason again being precisely that for the dual optimizer $\widehat{Y}_T(y)$ we may have $\mathbf{E}[\widehat{Y}_T(y)] < y$. Indeed, the validity of $u'(x) = \mathbf{E}[U'(\widehat{X}_T(x))]$ is tantamount to the validity of $y = \mathbf{E}[\widehat{Y}_T(y)]$.

CASE 2. $\mathrm{dom}(U) = \mathbb{R}$

We now pass to the case of a utility function U satisfying Assumption 1.2 Case 2 which is defined and finitely valued on all of $\mathbb{R}$. The reader should have in mind the exponential utility $U(x) = -e^{-\gamma x}$, for $\gamma > 0$, as the typical example.

We want to obtain a result analogous to Theorem 3.15 also in this setting. Roughly speaking, we get the same theorem, but the sets C and D considered above have to be chosen in a somewhat different way, as the optimal portfolio $\widehat{X}_T$ now may assume negative values too.

Firstly, we have to assume throughout the rest of this section that the semi-martingale S is *locally bounded*. The case of non locally bounded processes is not yet understood and waiting for future research.

Next we turn to the question; what is the proper definition of the set $C(x)$ of terminal values X_T dominated by a random variable $x + (H \cdot S)_T$, where H is an "allowed" trading strategy? On the one hand we cannot be too liberal in the choice of "allowed" trading strategies as we have to exclude doubling strategies and similar schemes. We therefore maintain the definition of the value function $u(x)$ unchanged

$$(3.70) \qquad u(x) = \sup_{H \in \mathcal{H}} \mathbf{E}\big[U\big(x + (H \cdot S)_T\big)\big], \quad x \in \mathbb{R},$$

where we still confine H to run through the set $\mathcal{H}$ of admissible trading strategies, i.e., such that the process $((H \cdot S)_t)_{0 \le t \le T}$ is uniformly bounded from below.

This notion makes good sense economically as it describes the strategies possible for an agent having a finite credit line.

On the other hand, in general, we have no chance to find the minimizer $\widehat{H}$ in (3.70) within the set of admissible strategies: already in the classical cases studied by Merton ([M69] and [M71] where, in particular, the case of exponential utility is solved for the Black-Scholes model) the optimal solution $x + (\widehat{H} \cdot S)_T$ to (3.70) is *not* uniformly bounded from below; this random variable typically assumes low values with very small probability, but its essential infimum typically is minus infinity.

In [S01] the following approach was used to cope with this difficulty: fix the utility function $U : \mathbb{R} \to \mathbb{R}$ and first define the set $C_U^b(x)$ to consist of all random variables G_T dominated by $x + (H \cdot S)_T$, for some *admissible* trading strategy H and such that $\mathbf{E}[U(G_T)]$ makes sense:

(3.71) $C_U^b(x) = \{ G_T \in L^0(\Omega, \mathcal{F}_T, \mathbf{P}) :$ there is H admissible s.t.

(3.72) $G_T \leq x + (H \cdot S)_T$ and $\mathbf{E}[|U(G_T)|] < \infty \}$.

Next we define $C_U(x)$ as the set of $\mathbb{R} \cup \{+\infty\}$-valued random variables X_T such that $U(X_T)$ can be approximated by $U(G_T)$ in the norm of $L^1(\mathbf{P})$, when G_T runs through $C_U^b(x)$:

(3.73) $C_U(x) = \{ X_T \in L^0(\Omega, \mathcal{F}_T, \mathbf{P}; \mathbb{R} \cup \{+\infty\}) : U(X_T)$ is in

(3.74) $L^1(P)$-closure of $\{ U(G_T) : G_T \in C_U^b(x) \} \}$.

The optimization problem (3.70) now reads

(3.75) $u(x) = \sup_{X_T \in C_U(x)} \mathbf{E}[U(X_T)], \quad x \in \mathbb{R}$.

The set $C_U(x)$ was chosen in such a way that the value functions $u(x)$ defined in (3.70) and (3.75) coincide; but now we have much better chances to find the maximizer to (3.75) in the set $C_U(x)$.

Two features of the definition of $C_U(x)$ merit some comment: firstly, we have allowed $X_T \in C_U(x)$ to attain the value $+\infty$; indeed, in the case when $U(\infty) < \infty$ (e.g., the case of exponential utility), this is natural, as the set $\{ U(X_T) : X_T \in C_U(x) \}$ should equal the $L^1(\mathbf{P})$-closure of the set $\{ U(G_T) : G_T \in C_U^b(x) \}$. But we shall see that – under appropriate assumptions – the optimizer $\widehat{X}_T$, which we are going to find in $C_U(x)$, will almost surely be finite.

Secondly, the elements X_T of $C_U(x)$ are only *random variables* and, at this stage, they are not related to a *process* of the form $x + (H \cdot S)$. Of course, we finally want to find for each $X_T \in C_U(x)$, or at least for the optimizer $\widehat{X}_T$, a predictable, S-integrable process H having "allowable" properties (in order to exclude doubling strategies) and such that $X_T \leq x + (H \cdot S)_T$. We shall prove later that – under appropriate assumptions – this is possible and give a precise meaning to the word "allowable".

After having specified the proper domain $C_U(x)$ for the primal optimization problem (3.75), we now pass to the question of finding the proper domain for the dual optimization problem. Here we find a pleasant surprise: contrary to Case 1 above, where we had to pass from the set $\mathcal{M}^a(S)$ to its closed, solid hull D, it turns out that, in the present Case 2, the dual optimizer always lies in $\mathcal{M}^a(S)$. This fact was first proved by F. Bellini and M. Fritelli ([BF02]).

We now can state the main result of [S01]:

THEOREM 3.19 [S01, Theorem 2.2]. *Let the locally bounded semi-martingale $S = (S_t)_{0 \le t \le T}$ and the utility function U satisfy Assumptions 1.1, 1.2 Case 2 and 1.3; suppose in addition that U has reasonable asymptotic elasticity. Define*

$$(3.76) \qquad u(x) = \sup_{X_T \in C_U(x)} \mathbf{E}[U(X_T)], \quad v(y) = \inf_{\mathbf{Q} \in \mathcal{M}^a(S)} \mathbf{E}\left[V\left(y\frac{d\mathbf{Q}}{d\mathbf{P}}\right)\right].$$

Then we have:
(i) *The value functions $u(x)$ and $v(y)$ are conjugate; they are continuously differentiable, strictly concave (resp. convex) on $\mathbb{R}$ (resp. on $]0, \infty[$) and satisfy*

$$(3.77) \qquad u'(-\infty) = -v'(0) = v'(\infty) = \infty, \quad u'(\infty) = 0.$$

(ii) *The optimizers $\widehat{X}_T(x)$ and $\widehat{\mathbf{Q}}(y)$ in (3.76) exist, are unique and satisfy*

$$(3.78) \qquad \widehat{X}_T(x) = I\left(y\frac{d\widehat{\mathbf{Q}}(y)}{d\mathbf{P}}\right), \quad y\frac{d\widehat{\mathbf{Q}}(y)}{d\mathbf{P}} = U'(\widehat{X}_T(x)),$$

where $x \in \mathbb{R}$ and $y > 0$ are related via $u'(x) = y$ or equivalently $x = -v'(y)$.
(iii) *We have the following relations between u', v' and $\widehat{X}, \widehat{\mathbf{Q}}$ respectively*:

$$(3.79) \qquad u'(x) = \mathbf{E_P}[U'(\widehat{X}_T(x))], \quad v'(y) = \mathbf{E}_{\widehat{\mathbf{Q}}}\left[V'\left(y\frac{d\widehat{\mathbf{Q}}(y)}{d\mathbf{P}}\right)\right]$$

$$(3.80) \qquad \begin{aligned} xu'(x) &= \mathbf{E_P}[\widehat{X}_T(x)U'(\widehat{X}_T(x))], \\ yv'(y) &= \mathbf{E_P}\left[y\frac{d\widehat{\mathbf{Q}}(y)}{d\mathbf{P}}V'\left(y\frac{d\widehat{\mathbf{Q}}(y)}{d\mathbf{P}}\right)\right]. \end{aligned}$$

(iv) *If $\widehat{\mathbf{Q}}(y) \in \mathcal{M}^e(S)$ and $x = -v'(y)$, then $\widehat{X}_T(x)$ equals the terminal value of a process of the form $\widehat{X}_t(x) = x + (H \cdot S)_t$, where H is predictable and S-integrable, and such that $\widehat{X}$ is a uniformly integrable martingale under $\widehat{\mathbf{Q}}(y)$.*

We refer to [S01] for a proof of this theorem and further related results. We cannot go into the technicalities here, but a few comments on the proof of the above theorem are in order: the technique is to reduce Case 2 to Case 1 by approximating the utility function $U : \mathbb{R} \to \mathbb{R}$ by a sequence $(U^{(n)})_{n=1}^{\infty}$ of utility functions $U^{(n)} : \mathbb{R} \to \mathbb{R} \cup \{-\infty\}$ such that $U^{(n)}$ coincides with U on $[-n, \infty[$ and equals $-\infty$ on $]-\infty, -(n+1)]$. For fixed initial endowment $x \in \mathbb{R}$, we then apply Theorem 3.15 to find for each $U^{(n)}$ the saddle-point $(\widehat{X}_T^{(n)}(x), \widehat{Y}_T^{(n)}(\widehat{y}_n)) \in C_U^b(x) \times D(\widehat{y}_n)$; finally we show that this sequence converges to some $(\widehat{X}_T(x), \widehat{y}\widehat{Q}_T) \in C_U(x) \times \widehat{y}\mathcal{M}^a(S)$, which then is shown to be the saddle-point for the present problem. The details of this construction are rather technical and lengthy (see [S01]).

We have assumed in item (iv) that $\widehat{Q}(y)$ is equivalent to $\mathbf{P}$ and left open the case when $\widehat{Q}(y)$ is only absolutely continuous to $\mathbf{P}$. F. Bellini and M. Fritelli have observed ([BF02]) that, in the case $U(\infty) = \infty$ (or, equivalently, $V(0) = \infty$), it follows from (3.76) that $\widehat{Q}(y)$ is equivalent to $\mathbf{P}$. But there are also other important cases where we can assert that $\widehat{Q}(y)$ is equivalent to $\mathbf{P}$: for example, for the case of the exponential utility $U(x) = -e^{-\gamma x}$, in which case the dual optimization becomes the problem of finding $\widehat{Q} \in \mathcal{M}^a(S)$ minimizing the relative entropy with respect $\mathbf{P}$, it follows from the work of Csiszar [C75] (compare also [R84], [F00], [GR01]) that the dual optimizer $\widehat{Q}(y)$ is equivalent to $\mathbf{P}$, provided only that there is at least one $\mathbf{Q} \in \mathcal{M}^e(S)$ with finite relative entropy.

Under the condition $\widehat{Q}(y) \in \mathcal{M}^e(S)$, item (iv) tells us that the optimizer $\widehat{X}_T \in C_U(x)$ is almost surely finite and equals the terminal value of a process $x + (H \cdot S)$, which is a uniformly integrable martingale under $\widehat{Q}(y)$; this property qualifies H to be a "allowable", as it certainly excludes doubling strategies and related schemes. One may turn the point of view around and take this as the *definition* of the "allowable" trading strategies; this was done in [DGRSSS02] for the case of exponential utility, where this approach is thoroughly studied and some other definitions of "allowable" trading strategies, over which the primal problem may be optimized, are also investigated. Further results on these lines were obtained in [KS02] for the case of exponential utility, and in [S03a] for general utility functions.

APPENDIX A

The bipolar theorem in L^0

We now present the non-locally convex version of the Bipolar theorem for $\langle L^0_+, L^0_+ \rangle$, was first proved by Brannath and Schachermayer [BS99]. The following proof is due to Michael Meyer (private communication).

We define the duality in $\langle L^0_+, L^0_+ \rangle$ given by $\langle f, g \rangle = \mathbf{E}[fg] \in [0, \infty]$. For a subset $C \subset L^0_+$ define the *polar* C^0 of C as

(A.1) $$C^0 = \{f \in L^0_+ : \mathbf{E_P}[fg] \leq 1, \forall g \in C\}$$

A subset $C \subset L^0_+$ will be called solid if $g \in C$, $h \in L^0$ and $0 \leq h \leq g$ implies that $h \in C$. Note that the polar C^0 of C is closed with respect to the topology of convergence in probability, convex and solid.

PROPOSITION A.1. *Assume that the set $C \subset L^0_+$ is nonempty, closed in probability, convex and solid. Then $C = C^{00}$.*

PROOF. By definition of the polar set we have $C \subset C^{00}$. Hence, it will thus suffice to show that if $f \notin C$ then $f \notin C^{00}$. Assume that $f \notin C$. Since f is almost surely finitely valued, we have $\mathbf{P}(f \neq f \wedge n) \to 0$ and so $f \wedge n \to f$ in probability, as $n \to \infty$. Since $f \notin C$ and C is closed in probability, we must have $f \wedge n \notin C$, for some $n \geq 1$. It will now suffice to show that $f \wedge n \notin C^{00}$, since the solidity of C^{00} then implies that $f \notin C^{00}$. This shows that we may assume that f is bounded. Let C_b denote the family of all bounded functions in C and let K be the closed convex hull

$$K = \overline{co}(C_b - L^1_+) \subset L^1$$

the closure being taken in the norm of L^1. If h is in the convex hull $co(C_b - L^1_+)$, then h_+ is dominated by a function in C and by the solidity of C we have that $h_+ \in C$. In short

$$h \in co(C_b - L^1_+) \to h_+ \in C.$$

Since the map $h \mapsto h_+$ is continuous in L^1-norm, convergence in L^1-norm implies convergence in probability and C is closed in probability, it follows that

$$h \in K = \overline{co}(C_b - L^1_+) \subset L^1 \to h_+ \in C.$$

Since $f = f_+$ and $f \notin C$ it follows that $f \notin K$. Using the convex separation theorem in L^1 and the canonical duality $(L^1)^* = L^\infty$, we obtain $g \in L^\infty$ such that

$$\mathbf{E}(gh) \leq 1, \quad \forall h \in K \quad \text{and} \quad \mathbf{E}[gf] > 1.$$

Since C is nonempty and solid, we have $0 \in C_b$ and thus $-L^1_+ \subset K$. Consequently $\mathbf{E}[gh] \leq 1$, for all $h \in -L^1_+$, and so $g \geq 0$ almost surely. Moreover $\mathbf{E}[gh] \leq 1$, for all $h \in C_b$ and consequently for all $h \in C$ (if $h \in C$, then $h \wedge n \in C_b$, for all $n \geq 1$, by solidity of C, and $h \wedge n \to h$ as $n \to \infty$). Thus $g \in C^0$ and $\mathbf{E}[gf] > 1$. It follows that $f \notin C^{00}$.

THEOREM A.2 (Bipolar theorem). *Let $C \subset L^0_+$ be non-empty. Then the bipolar C^{00} is the closed convex solid hull of C in L^0_+ (closure in the topology of convergence in probability).*

PROOF. For subsets $A, B \subset L^0_+$ we have $A \subset B \Rightarrow B^0 \subset A^0 \Rightarrow A^{00} \subset B^{00}$, by definition of the polar set. The bipolar C^{00} is convex, solid and closed and contains the set C. Thus the closed, convex solid hull $\widetilde{C}$ of C satisfies $\widetilde{C} \subset C^{00}$. On the other hand $C \subset \widetilde{C}$ and so $C^{00} \subset (\widetilde{C})^{00} = \widetilde{C}$, where the last equality uses. It follows that $C^{00} = \widetilde{C}$. $\square$

APPENDIX B

Asymptotic elasticity and asymptotic relative risk aversion

Let $U : \mathbb{R} \to \mathbb{R} \cup \{\infty\}$ be a utility function satisfying our usual regularity hypothesis (Assumtpion 1.2). We say that U satisfies $AE_{+\infty}(U) < 1$ if

$$(B.1) \qquad AE_{+\infty}(U) = \limsup_{x \to \infty} \frac{xU'(x)}{U(x)} < 1.$$

If U is twice differentiable, we say that the asymptotic relative risk aversion is positive if

$$ARRA_{+\infty}(U) = \lim_{x \to \infty} \left(-\frac{xU''(x)}{U(x)} \right)$$

exists and is strictly positive.

P. Guasoni observed that the two conditions are obviously equivalent if de l'Hôpital's formula is applicable (we then suppose in particular that the upper limit in (2.10) is a limit. Indeed, a formal application of this rule yields

$$(B.2) \qquad \lim_{x \to \infty} \frac{xU'(x)}{U(x)} = \lim_{x \to \infty} \frac{xU''(x) + U'(x)}{U'(x)} = 1 - \lim_{x \to \infty} \left(-\frac{xU''(x)}{U'(x)} \right)$$

which readily yields the claimed equivalence. Passing from this formal argument to a well-defined setting we may formulate the following:

PROPOSITION B.1. *Suppose that U, apart from the usual hypotheses, is twice differentiable and that the subsequent limit exists (possibly taking the value $+\infty$).*

$$(B.2a) \qquad ARRA_{+\infty}(U) = \lim_{x \to \infty} \left(-\frac{xU''(x)}{U'(x)} \right).$$

Then the limit

$$(B.2b) \qquad AE_{+\infty}(U) = \lim_{x \to \infty} \frac{xU'(x)}{U(x)}$$

exists too, and $AE_{+\infty}(U) < 1$ iff $ARRA_{+\infty}(U) > 0$.

PROOF. For $x \geq 1$ we may write

$$xU'(x) = 1U'(1) + \int_1^x (zU'(z))'dz = U'(1) + \int_1^x \left[U'(z) + zU''(z) \right] dz.$$

We distinguish two cases:

CASE 1. $U(\infty) < \infty$ or, equivalently $\int_0^\infty U'(z)dz < \infty$.

In this case it follows from the monotonicity of U' that $\lim_{x \to \infty} xU'(x) = 0$. Hence, if $U(\infty) \neq 0$ we have:

$$\lim_{x \to \infty} \frac{xU'(x)}{U(x)} = 0.$$

If $U(\infty) = 0$, by hypothesis (A2a) we are entitled to apply de l'Hôpital, and

$$\lim_{x \to \infty} \frac{xU'(x)}{U(x)} = \lim_{x \to \infty} \left(1 - \frac{xU''(x)}{U'(x)} \right).$$

Hence we have $\mathrm{AE}_{+\infty}(U) = 1 - \mathrm{ARRA}_{+\infty}(U)$. Noting the obvious fact (Lemma 3.2) that $U(\infty) = 0$ implies that $\mathrm{AE}_{+\infty}(U) \leq 0$ and that $\mathrm{ARRA}_{+\infty}(U)$ does not change by adding a constant to U, we also obtain that $U(\infty) < \infty$ implies $\mathrm{AE}_{+\infty}(U) \leq 0$ and $\mathrm{ARRA}_{+\infty}(U) \geq 1$. This proves the claim for the case $U(\infty) < \infty$.

CASE 2. $U(\infty) = \infty$ or, equivalently $\int_1^\infty U'(z)dz = \infty$.

Denoting $a = \mathrm{ARRA}_{+\infty}(U) \in [0, \infty]$ we then have by (A2a):

$$\lim_{x \to \infty} \frac{\int_1^x zU''(z)dz}{\int_1^x U'(z)dz} = \lim_{x \to \infty} \frac{\int_1^x zU''(z)dz}{U(x)} = -a.$$

Hence:

$$\lim_{x \to \infty} \frac{xU'(x)}{U(x)} = \lim_{x \to \infty} \frac{\int_1^x \left(U'(z) + zU''(z) \right) dz}{\int_1^x U'(z)dz} = 1 - a. \qquad \square$$

REMARK B.2.(a) When does the formula

(B.3) $\mathrm{AE}_{+\infty}(U) = 1 - \mathrm{ARRA}_{+\infty}(U)$

hold true? We have seen in the above argument that this is the case when either $U(\infty) = \infty$ or $U(\infty) = 0$. But in the case $U(\infty) \in \mathbb{R} \setminus \{0\}$ there is no reason for this relation to hold true. In fact, $\mathrm{AE}_{+\infty}(U)$ may change when shifting $U(x)$ by a constant, while $\mathrm{ARRA}_{+\infty}(U)$ is not affected. Consider for instance the utility functions $U(x) = -x^{-\alpha} + c$ for $\alpha > 0$, for which $\text{ARRA}_{+\infty}(U) = \alpha + 1$, while $\text{AE}_{+\infty}(U)$ depends on c.

In Case 2, one might try to drop the assumption in (B.2a) that $\lim_{x \to \infty} \frac{xU''(x)}{U'(x)}$ exists and define

$$\text{ARRA}_{+\infty}(U) = \liminf_{x \to \infty} \left(-\frac{xU''(x)}{U'(x)} \right)$$

but then the assertion of the proposition is not valid any more, as can be seen from the following observation: let U be any utility function. To avoid trivialities we suppose $U(\infty) > 0$. If $\overline{U}$ is any other utility function such that $\overline{U}(n) = U(n)$ for $n \in \mathbb{N}$, we have :

$$\text{AE}_{+\infty}(U) = \text{AE}_{+\infty}(\overline{U}).$$

Indeed, a little picture reveals that the concavity of $\overline{U}$ implies that

$$U'(n+1) \le \overline{U}'(n) \le U'(n-1)$$

and it is easy to see that

$$\limsup_{\substack{x \to \infty \\ x \in R}} \frac{xU'(x)}{U(x)} = \limsup_{\substack{n \to \infty \\ n \in \mathbb{N}}} \frac{nU'(n)}{U(n)}.$$

Whence:

$$\text{AE}_{+\infty}(U) = \limsup_{n \to \infty} \frac{nU'(n+1)}{U(n)} \le \limsup_{n \to \infty} \frac{n\overline{U}'(n)}{\overline{U}(n)}$$
$$\le \limsup_{n \to \infty} \frac{nU'(n-1)}{U(n)} \le \text{AE}_{+\infty}(U).$$

Hence modifying $U(x)$ between the integer points $U(n)$, while leaving it monotone, differentiable, and strictly concave, does not affect $\text{AE}_{\infty}(U)$. On the other hand, it is rather obvious that we can do such modifications such that the behavior of $\overline{U}''(x)$ oscillates in an arbitrarily wild way between $-\infty$ and 0 (assuming that $\overline{U}$ is twice differentiable). Hence replacing the limit in the definition of $\text{ARRA}_{+\infty}(U)$ in (B.2a) by a lower or an upper limit cannot determine the value of $\text{AE}_{\infty}(U)$.

We only have the following more modest result, which we stat without proof.

PROPOSITION B.3. *If* $\lim\inf_{x \to \infty} \left(-\frac{xU''(x)}{U'(x)} \right) = a > 0$ *then* $\text{AE}_{+\infty}(U) < 1$ *and, in fact,* $\text{AE}_{+\infty}(U) \le (1-a)_+$.

If $\text{AE}_{+\infty}(U) = b < 1$, *then* $\limsup_{x \to \infty} \left(-\frac{xU''(x)}{U'(x)} \right) \ge (1-b) \wedge 1 > 0$.

REMARK B.4. Let us now pass to the situation at $-\infty$: this case is easier as it follows from Assumption 1.2 Case 2 that $\lim_{x\to\infty} \frac{xU'(x)}{U(x)}$ always is of the form $\frac{-\infty}{-\infty}$. Hence assuming that:

$$\mathrm{ARRA}_{-\infty}(U) = \lim_{x\to\infty} -\frac{xU''(x)}{U'(x)} \in [0, \infty]$$

exists, we may again apply de l'Hôpital's rule to obtain $\mathrm{AE}_{-\infty}(U) = 1 - \mathrm{ARRA}_{-\infty}(U)$. The other considerations on the non-existence of the limit also carry over.

Summing up, if we stick to utility functions such that $-\frac{xU''(x)}{U'(x)}$ is well-defined and converges, as x tends to $+\infty$ or $-\infty$, the reasonable asymptotic elasticity may equivalently be described in terms of the asymptotic relative risk aversion.

However, the notion of reasonable asymptotic elasticity also carries over to general utility functions and gives in this general class a necessary and sufficient condition for a good utility maximization theory. It is not possible to formulate analogous necessary and sufficient conditions in terms of ARRA.

Bibliography

[AS94]   J. P. ANSEL – C. STRICKER, *Couverture des actifs contingents et prix maximum*, Ann. Inst. Henri Poincaré **30** (1994), 303-315.

[A97]   P. ARTZNER, "On the numeraire portfolio", Mathematics of Derivative Securities, M. Dempster and S. Pliska, eds., Cambridge University Press, 1997, 53-60.

[B01]   D. BECHERER, *The numeraire portfolio for unbounded semimartingales*, Finance and Stochastics **5** no. 3 (2001), 327-341.

[BF02]   F. BELLINI – M. FRITELLI, *On the existence of minimax martingale measures*, Mathematical Finance **12** no. 1 (2002), 1-21.

[BS99]   W. BRANNATH – W. SCHACHERMAYER, *A Bipolar Theorem for Subsets of $L_+^0(\Omega, \mathcal{F}, P)$*, Séminaire de Probabilités **XXXIII** (1999), 349-354.

[CH00]   P. COLLIN-DUFRESNE – J.-N. HUGGONNIER, "Utility-based pricing of contingent claims subject to counterparty credit risk", Working paper GSIA & Department of Mathematics, Carnegie Mellon University, 2000.

[CH89]   J. C. COX – C. F. HUANG, *Optimal consumption and portfolio policies when asset prices follow a diffusion process*, Journal of Economic Theory **49** (1989), 33-83.

[CH91]   J. C. COX – C. F. HUANG, *A variational problem arising in financial economics*, Jorunal of Mathematical Economics **20** no. 5 (1991), 465-487.

[C75]   I. CSISZAR, *I-Divergence Geometry of Probability Distributions and Minimization Problems*, Annals of Probability **3** no. 1 (1975), 146-158.

[C00]   J. CVITANIĆ, *Minimizing expected loss of hedging in incomplete and constrained markets*, SIAM Journal on Control and Optimization **38** no. 4 (2000), 1050-1066 (electronic).

[CK96]   J. CVITANIC – I. KARATZAS, *Hedging and portfolio optimization under transaction costs: A martingale approach*, Mathematical Finance **6** no. 2 (1996), 133-165.

[CSW01]   J. CVITANIC – W. SCHACHERMAYER – H. WANG, *Utility Maximization in Incomplete Markets with Random Endowment*, Finance and Stochastics **5** no. 2 (2001), 259-272.

[CW01]   J. CVITANIC – H. WANG, *On optimal terminal wealth under transaction costs*, Jorunal of Mathematical Economics **35** no. 2 (2001), 223-231.

[DMW90]   R. C. DALANG – A. MORTON – W. WILLINGER, *Equivalent martingale measures and no-arbitrage in stochastic*, Stochastics and Stochastics Reports **29** (1990), 185-201.

[D97]   M. DAVIS, "Option pricing in incomplete markets", Mathematics of Derivative Securities, eds. M.A.H. Dempster and S.R. Pliska, Cambridge University Press, 1997, 216-226.

[D00]   M. DAVIS, "Option valuation and hedging with basis risk", System theory: modeling, analysis and control (Cambridge, MA, 1999), Kluwer Internat. Ser. Engrg. Comput. Sci., vol. 518, Kluwer Acad. Publ., Boston, MA, 2000, 245-254.

[DPT01]   G. DEELSTRA – H. PHAM – N. TOUZI, *Dual formulation of the utility maximisation problem under transaction costs*, Annals of Applied Probability **11** no. 4 (2001), 1353-1383.

[DGRSSS02] F. DELBAEN – P. GRANDITS, T. RHEINLNDER – D. SAMPERI – M. SCHWEIZER – C. STRICKER, *Exponential hedging and entropic penalties*, Mathematical Finance **12** no. 2 (2002), 99-123.

[DS94]   F. DELBAEN – W. SCHACHERMAYER, *A General Version of the Fundamental Theorem of Asset Pricing*, Math. Annalen **300** (1994), 463-520.

[DS95]   F. DELBAEN – W. SCHACHERMAYER, *The No-Arbitrage Property under a change of numéraire*, Stochastics and Stochastic Reports **53** (1995), 213-226.

[DS98]   F. DELBAEN – W. SCHACHERMAYER, *The Fundamental Theorem of Asset Pricing for Unbounded Stochastic Processes*, Mathematische Annalen **312** (1998), 215-250.

[DS98a]   F. DELBAEN – W. SCHACHERMAYER, *A Simple Counter-example to Several Problems in the Theory of Asset Pricing, which arises in many incomplete markets*, Mathematical Finance **8** (1998), 1-12.

[DS99]   F. DELBAEN – W. SCHACHERMAYER, *A Compactness Principle for Bounded Sequences of Martingales with Applications*, Proceedings of the Seminar of Stochastic Analysis, Random Fields and Applications, Progress in Probability **45** (1999), 137-173.

[ET76]   I. EKELAND – R. TEMAM, "Convex Analysis and Variational Problems", North Holland, 1976.

[E80]   M. EMERY, *Compensation de processus à variation finie non localement intégrables*, Séminaire de Probabilités XIV, Springer Lecture Notes in Mathematics **784** (1980), 152-160.

[F90]   L. P. FOLDES, *Conditions for optimality in the infinite-horizon portfolio-cum-savings problem with semimartingale investments*, Stochastics and Stochastics Report **29** (1990), 133-171.

[FL00]   H. FÖLLMER – P. LEUKERT, *Efficient Hedging: Cost versus Shortfall Risk*, Finance and Stochastics **4** no. 2 (2000), 117-146.

[FS91]   H. FÖLLMER – M. SCHWEIZER, "Hedging of contingent claims under incomplete information", Applied Stochastic Analysis, Stochastic Monographs, M.H.A. Davis and R.J. Elliott, eds., Gordon and Breach, London, New York, Vol. 5, 1991, 389-414.

[F00]   M. FRITELLI, *The minimal entropy martingale measure and the valuation problem in incomplete markets*, Mathematical Finance **10** no. 1 (2000), 39-52.

[GK00]   T. GOLL – J. KALLSEN, *Optimal portfolios for logarithmic utility*, Stochastic Processes and Their Applications **89** (2000), 31-48.

[GR01]   T. GOLL – L. RSCHENDORF, *Minimax and minimal distance martingale measures and their relationship to portfolio optimization*, Finance and Stochastics **5** no. 4 (2001), 557-581.

[GS04]   P. GUASONI – W. SCHACHERMAYER, *Necessary Conditions for the Existence of Utility Maximizing Strategies under Transaction Costs*, Preprint (2003), (19 pages).

[HP81]   J. M. HARRISON – S. R. PLISKA, *Martingales and stochastic integrals in the theory of continuous trading*, Stochastic Processes and Applications **11** (1981), 215-260.

[HP91a] H. HE – N. D. PEARSON, *Consumption and Portfolio Policies with Incomplete Markets and Short-Sale Constraints: The Finite-Dimensional Case*, Mathematical Finance **1** (1991), 1-10.

[HP91b] H. HE – N.D. PEARSON, *Consumption and Portfolio Policies with Incomplete Markets and Short-Sale Constraints: The Infinite-Dimensional Case*, Journal of Economic Theory **54** (1991), 239-250.

[HN89]   S. D. HODGES – A. NEUBERGER, *Optimal replication of contingent claims under transaction costs*, Review of Futures Markets **8** (1989), 222-239.

[HL88]   C.-F. HUANG – R. H. LITZENBERGER, "Foundations for Financial Economics", North-Holland Publishing Co., New York, 1988.

[HK02]   J. HUGONNIER – D. KRAMKOV, *Optimal investment with random endowments in incomplete markets*, to appear in The Annals of Applied Probability, (2002).

[J92]   S. D. JACKA, *A martingale representation result and an application to incomplete financial markets*, Mathematical Finance **2** (1992), 239-250.

[KS01]   Yu. M. KABANOV – CH. STRICKER, *A teachers' note on no-arbitrage criteria*, Séminaire de Probabilités XXXV, Springer Lecture Notes in Mathematics **1755** (2001), 149-152.

[KS02]   Yu. M. KABANOV – C. STRICKER, *On the optimal portfolio for the exponential utility maximization: remarks to the six-author paper*, Mathematical Finance **12** no. 2 (2002), 125-134.

[KP65]   M. Ĭ. KADEC' – A. PELČZYN' SKI, *Basic sequences, bi-orthogonal systems and norming sets in Banach and Frechet spaces*, Studia Mathematica **25** (1965), 297-323.

[K00]   J. KALLSEN, *Optimal portfolios for exponential Lévy processes*, Mathematical Methods of Operation Research **51** no. 3 (2000), 357-374.

[KLS87]   I. KARATZAS – J. P. LEHOCZKY – S. E. SHREVE, *Optimal portfolio and consumption decisions for a "small investo" on a finite horizon*, SIAM Journal of Control and Optimization **25** (1987), 1557-1586.

[KLSX91]   I. KARATZAS – J. P. LEHOCZKY – S. E. SHREVE – G. L. XU, *Martingale and duality methods for utility maximization in an incomplete market*, SIAM Journal of Control and Optimization **29** (1991), 702-730.

[KJ98]   N. EL KAROUI – M. JEANBLANC, *Optimization of consumptions with labor income*, Finance and Stochastics **4** (1998), 409-440.

[KQ95]   N. EL KAROUI – M.-C. QUENEZ, *Dynamic programming and pricing of contingent claims in an incomplete market*, SIAM Journal on Control and Optimization **33** (1995), 29-66.

[KR00]   N. EL KAROUI – R. ROUGE, *Pricing via utility maximization and entropy*, Mathematical Finance **10** no. 2 (2000), 259-276.

[Kom67]   J. KOMLOS, *A generalization of a problem of Steinhaus*, Acta Math. Sci. Hung. **18** (1967), 217-229.

[KS99]   D. KRAMKOV – W. SCHACHERMAYER, *The Asymptotic Elasticity of Utility Functions and Optimal Investment in Incomplete Markets*, Annals of Applied Probability **9** no. 3 (1999), 904-950.

[K81]   D. M. KREPS, *Arbitrage and equilibrium in economies with infinitely many commodities*, Journal of Mathematical Economics **8** (1981), 15-35.

[L00]   P. LAKNER, *Portfolio Optimization with an Insurance Constraint*, preprint of the NYU, Dept. of Statistics and Operation Research, (2000).

[L90]   J. B. LONG, *The numeraire portfolio*, Journal of Financial Economics **26** (1990), 29-69.

[M69]   R. C. MERTON, *Lifetime portfolio selection under uncertainty: the continuous-time model*, Rev. Econom. Statist. **51** (1969), 247-257.

[M71]   R. C. MERTON, *Optimum consumption and portfolio rules in a continuous-time model*, Journal of Economic Theory **3** (1971), 373-413.

[M90]   R. C. MERTON, "Continuous-Time Finance", Basil Blackwell, Oxford, 1990.

[P86]   S. R. PLISKA, *A stochastic calculus model of continuous trading: optimal portfolios*, Math. Oper. Res. **11** (1986), 371-382.

[R70]   R. T. ROCKAFELLAR, "Convex Analysis", Princeton University Press, Princeton, New Jersey, 1970.

[R84]   L. RÜSCHENDORF, *On the minimum discrimination information theorem*, Statistics & Decisions Supplement Issue **1** (1984), 263-283.

[S69]   P. A. SAMUELSON, *Lifetime portfolio selection by dynamic stochastic programming*, Rev. Econom. Statist. **51** (1969), 239-246.

[S01]   W. SCHACHERMAYER, *Optimal Investment in Incomplete Markets when Wealth may Become Negative*, Annals of Applied Probability **11** no. 3 (2001), 694-734.

[S01a]   W. SCHACHERMAYER, *Optimal Investment in Incomplete Financial Markets*, Mathematical Finance: Bachelier Congress 2000 (H. Geman, D. Madan, St.R. Pliska, T. Vorst, editors), Springer, 2001, 427-462.

[S03]   W. SCHACHERMAYER, "Introduction to the Mathematics of Financial Markets", In: S. Albeverio, W. Schachermayer, M. Talagrand: Lecture Notes in Mathematics 1816 - Lectures on Probability Theory and Statistics, Saint-Flour summer school 2000 (Pierre Bernard, editor), Springer Verlag, Heidelberg, 2003, 111-177.

[S03a]   W. SCHACHERMAYER, *A Super-Martingale Property of the Optimal Portfolio Process*, Finance and Stochastics **7** no. 4 (2003), 433-456.

[S04]   W. SCHACHERMAYER, *The Fundamental Theorem of Asset Pricing under Proportional Transaction Costs in Finite Discrete Time*, Mathematical Finance **14** no. 1 (2004), 19-48.

[S66]   H. H. SCHÄFER, "Topological Vector Spaces", Graduate Texts in Mathematics, 1966.

[St85]   H. STRASSER, *Mathematical theory of statistics: statistical experiments and asymptotic decision theory*, De Gruyter studies in mathematics **7** (1985).

PUBBLICAZIONI DELLA CLASSE DI SCIENZE DELLA SCUOLA NORMALE SUPERIORE

QUADERNI

1. DE GIORGI E., COLOMBINI F., PICCININI L.C.: *Frontiere orientate di misura minima e questioni collegate.*
2. MIRANDA C.: *Su alcuni problemi di geometria differenziale in grande per gli ovaloidi.*
3. PRODI G., AMBROSETTI A.: *Analisi non lineare.*
4. MIRANDA C.: *Problemi in analisi funzionale* (ristampa).
5. TODOROV I.T., MINTCHEV M., PETKOVA V.B.: *Conformal Invariance in Quantum Field Theory.*
6. ANDREOTTI A., NACINOVICH M.: *Analytic Convexity and the Principle of Phragmén-Lindelöf.*
7. CAMPANATO S.: *Sistemi ellittici in forma divergenza. Regolarità all'interno.*
8. TOPICS IN FUNCTIONAL ANALYSIS: *Contributors:* F. STROCCHI, E. ZARANTONELLO, E. DE GIORGI, G. DAL MASO, L. MODICA.
9. LETTA G.: *Martingales et intégration stochastique.*
10. OLD AND NEW PROBLEMS IN FUNDAMENTAL PHYSICS: *Meeting in honour of* GIAN CARLO WICK.
11. INTERACTION OF RADIATION WITH MATTER: *A Volume in honour of* ADRIANO GOZZINI.
12. MÉTIVIER M.: *Stochastic Partial Differential Equations in Infinite Dimensional Spaces.*
13. SYMMETRY IN NATURE: *A Volume in honour of* LUIGI A. RADICATI DI BROZOLO.
14. NONLINEAR ANALYSIS: *A Tribute in honour of* GIOVANNI PRODI.
15. LAURENT-THIÉBAUT C., LEITERER J.: *Andreotti-Grauert Theory on Real Hypersurfaces.*
16. ZABCZYK J.: *Chance and Decision. Stochastic Control in Discrete Time.*
17. EKELAND I.: *Exterior Differential Calculus and Applications to Economic Theory.*
18. ELECTRONS AND PHOTONS IN SOLIDS: *A Volume in honour of* FRANCO BASSANI.
19. ZABCZYK J.: *Topics in Stochastic Processes.*
20. TOUZI N.: *Stochastic Control Problems, Viscosity Solutions and Application to Finance.*

CATTEDRA GALILEIANA

1. LIONS P.L.: *On Euler Equations and Statistical Physics.*
2. BJÖRK T.: *A Geometric View of the Term Structure of Interest Rates.*
3. DELBAEN F.: *Coherent Risk Measures.*
4. SCHACHERMAYER W.: *Portfolio Optimization in Incomplete Financial Markets.*

LEZIONI LAGRANGE

1. VOISIN C.: *Variations of Hodge Structure of Calabi-Yau Threefolds.*

LEZIONI FERMIANE

1. THOM R.: *Modèles mathématiques de la morphogénèse.*
2. AGMON S.: *Spectral Properties of Schrödinger Operators and Scattering Theory.*
3. ATIYAH M.F.: *Geometry of Yang-Mills Fields.*
4. KAC M.: *Integration in Function Spaces and Some of Its Applications.*
5. MOSER J.: *Integrable Hamiltonian Systems and Spectral Theory.*
6. KATO T.: *Abstract Differential Equations and Nonlinear Mixed Problems.*
7. FLEMING W.H.: *Controlled Markov Processes and Viscosity Solution of Nonlinear Evolution Equations.*
8. ARNOLD V.I.: *The Theory of Singularities and Its Applications.*

9. OSTRIKER J.P.: *Development of Larger-Scale Structure in the Universe.*
10. NOVIKOV S.P.: *Solitons and Geometry.*
11. CAFFARELLI L.A.: *The Obstacle Problem.*
12. CHEEGER J.: *Degeneration of Riemannian metrics under Ricci curvature bounds.*

PUBBLICAZIONI DEL CENTRO DI RICERCA MATEMATICA ENNIO DE GIORGI

1. DYNAMICAL SYSTEMS. Part I: *Hamiltonian Systems and Celestial Mechanics.*
2. DYNAMICAL SYSTEMS. Part II: *Topological, Geometrical and Ergodic Properties of Dynamics.*
3. MATEMATICA, CULTURA E SOCIETÀ *2003.*
4. RICORDANDO FRANCO CONTI.
5. N. KRYLOV: *Probabilistic Methods of Investigating Interior smoothness of Harmonic Functions Associated with Degenerate Elliptic Operators.*

ALTRE PUBBLICAZIONI

Proceedings of the Symposium on FRONTIER PROBLEMS IN HIGH ENERGY PHYSICS Pisa, June 1976

Proceedings of International Conferences on SEVERAL COMPLEX VARIABLES, Cortona, June 1976 and July 1977

Raccolta degli scritti dedicati a JEAN LERAY apparsi sugli Annali della Scuola Normale Superiore di Pisa

Raccolta degli scritti dedicati a HANS LEWY apparsi sugli Annali della Scuola Normale Superiore di Pisa

Indice degli articoli apparsi nelle Serie I, II e III degli Annali della Scuola Normale Superiore di Pisa (dal 1871 al 1973)

Indice degli articoli apparsi nella Serie IV degli Annali della Scuola Normale Superiore di Pisa (dal 1974 al 1990)

ANDREOTTI A.: *SELECTA vol. I, Geometria algebrica.*

ANDREOTTI A.: *SELECTA vol. II, Analisi complessa, Tomo I e II.*

ANDREOTTI A.: *SELECTA vol. III, Complessi di operatori differenziali.*

Fotocomposizione "CompoMat" Loc. Braccone, 02040 Configni (RI), Italy
Finito di stampare per conto della "CompoMat" dalla Nuova Grafica 86 nel novembre 2004